WITHDRAWN
IOWA STATE UNIVERSITY
LIBRARY

Texts in
Computational Science
and Engineering

4

Editors

Timothy J. Barth
Michael Griebel
David E. Keyes
Risto M. Nieminen
Dirk Roose
Tamar Schlick

Henry Gardner · Gabriele Manduchi

Design Patterns for e-Science

With 60 Figures

Henry Gardner
Computer Science, FEIT
College of Engineering and Computer Science
Australian National University
Canberra ACT 0200, Australia
email: henry.gardner@anu.edu.au

Gabriele Manduchi
Consorzio RFX
Associazione EURATOM-ENEA sulla Fusione
Corso Stati Uniti 4
35127 Padova, Italy
email: gabriele.manduchi@igi.cnr.it

Library of Congress Control Number: 2006940183

Mathematics Subject Classification (2000): 68N19, 68U01, 68U35

ISSN 1611-0994
ISBN-10 3-540-68088-8 Springer Berlin Heidelberg New York
ISBN-13 978-3-540-68088-8 Springer Berlin Heidelberg New York

This work is subject to copyright. All rights are reserved, whether the whole or part of the material is concerned, specifically the rights of translation, reprinting, reuse of illustrations, recitation, broadcasting, reproduction on microfilm or in any other way, and storage in data banks. Duplication of this publication or parts thereof is permitted only under the provisions of the German Copyright Law of September 9, 1965, in its current version, and permission for use must always be obtained from Springer. Violations are liable for prosecution under the German Copyright Law.

Springer is a part of Springer Science+Business Media
springeronline.com
© Springer-Verlag Berlin Heidelberg 2007
Printed in Germany

The use of general descriptive names, registered names, trademarks, etc. in this publication does not imply, even in the absence of a specific statement, that such names are exempt from the relevant protective laws and regulations and therefore free for general use.

Typesetting: by the author using a Springer T$_E$X macro package
Cover design: *WMXDesign*, Heidelberg
Production: VTEX, Vilnius

Printed on acid-free paper   SPIN: 11947356   46/3100/VTEX - 5 4 3 2 1 0

In Memory of Mark Jarnyk

Preface

This is a book about a code and about coding. The code is a case study which has been used to teach courses in e-Science at the Australian National University since 2001. Students learn advanced programming skills and techniques in the Java^{TM} language. Above all, they learn to apply useful object-oriented design patterns as they progressively refactor and enhance the software.

We think our case study, `EScope`, is as close to real life as you can get! It is a smaller version of a networked, graphical, waveform browser which is used in the control rooms of fusion energy experiments around the world. It is quintessential "e-Science" in the sense of e-Science being "computer science and information technology in the service of science". It is not, specifically, "Grid-enabled", but we develop it in a way that will facilitate its deployment onto the Grid. The standard version of `EScope` interfaces with a specialised database for waveforms, and related data, known as `MDSplus`. On the accompanying CD, we have provided you with software which will enable you to install `MDSplus`, `EScope` and sample data files onto Windows or Linux computers. There is much additional software including many versions of the case study as it gets built up and progressively refactored using design patterns. There will be a home web-site for this book which will contain up-to-date information about the software and other aspects of the case study.

If you journey with us, you will end up building a genuinely useful piece of software which you will be able to reuse for a number of applications: as a waveform browser for scientific, medical, commercial and stock-market data, as a front-end and back-end for scientific simulation codes, as a web-page "scraper" and so forth. More importantly, you will learn to live and breathe design patterns. We have been selective about the patterns which we have chosen to illustrate using our case study. The ones which we use really do work! We have described the patterns from a practitioner's standpoint and have tried to make them as simple and digestible as we can.

Few topics are more important than design patterns if you wish to be able to develop high-quality, object-oriented software systems.

Pedagogical Structure

As authors, we assume that you, our readers, might be scientists and engineers who have learnt "a bit of Java" and who wish to teach yourself enough to program Java (or C#^{TM} or other, similar languages) for your own work. You might be students who are converting to information technology from a science or engineering background or you might be information technology students who wish to learn about e-Science applications. You might even be computer science professionals who wish to learn more about e-Science from a requirements perspective. We think that this book is ideally configured for self-study. If you follow the text in, roughly, sequential order and you attempt to code each revision of the software before looking at our "answers", then you will pick up some good programming skills and you will really understand the patterns that we have used.

Substantial parts of the material in this book have been used to teach courses in e-Science since 2001. It is positioned as a second programming course for graduate, conversion students which is roughly equivalent to a second year course for information technology majors. But, in fact, the material can be used quite flexibly depending on the interests, and programming background, of students. The overall structure of a course would comprise two parts:

- **Part I:**
  Students learn relevant parts of the Java^{TM} API and build up a GUI interface to a server which contains waveform data. Chapters 1-6 of the book provide a number of (hopefully) entertaining examples and exercises which all work towards constructing this initial code, which we call `PreEScope`. We aim to provide enough background to enable students to build the software on their own. It is best if several of the steps in this part of the software process are completed by students in laboratory sessions. Additional, small laboratory exercises are suggested in order to inspire students to experiment creatively with new parts of the Java API. Several classes need to be provided as "black boxes" to connect and send information to the `MDSplus` database. Some of the graphics classes, particularly to draw and annotate the graph axes, would be best provided to students rather than being written from scratch. In the past, the step roughly equivalent to that from `PreEScope3` to `PreEScope4`, has been set as a programming assignment.

- **Part II:**
  After a theoretical introduction to software engineering and design patterns, students then proceed to refactor their code from the first part of the course. The refactoring steps are described in Chapters 8-19 of this book. There is much more material here than can be fitted into a single semester course together with Part I, so a judicious selection is needed. It is very important that students understand the initial refactoring into

domains using the facade pattern and they should get an opportunity to try this themselves in the laboratory. The use of the adapter and observer patterns to enable communication between domains should also be explained. But the detailed refactoring of the graphics domain following the treatment in Chapters 10-13 could be skipped over if a students had a greater interest in networking to multiple data-servers. Alternatively, a course which was more focussed on graphics could skip chapters 15-18. All of this depends on the interests and programming background of students.

Courses taught using the material in this book have been assessed using a mixture of assignments, a written examination and a practical examination. With the publication of this book, lecturers will find themselves in the situation that students can turn to the appendices, or load up the CD, and find the answers to many of the laboratory exercises and assignments! Here are some crafty ideas about how this material might be used for assessed laboratory exercises and assignments where it was important that students handed in original work:

1. For minor assignments, the starting-point version of the code could be modified "cosmetically" from that provided here. Small changes to the software will require an attention to the details of the patterns involved in its construction and refactoring.
2. For slightly larger assignments, the order that the refactoring steps are undertaken could be varied. For example, the decorator pattern could be introduced before the template pattern. For another example, the introduction of the observer pattern could be made earlier and its format might be simplified from our somewhat complicated treatment in Chapter14.
3. For large assignments, additional functionality could be provided. EScope is a very rich case study and there are many ways in which it can be further enhanced. (In fact, there is Enormous Scope for variations and extensions of this case study!) Examples on the graphics side are: individual waveforms from a group of waveforms could be dragged, the axes for a group of waveforms could change depending on which waveform was selected, a state pattern for mouse state could be implemented and so on. On the data server side, new text-oriented data servers could be implemented (perhaps using XML), a relational database index to MDSplus data could be implemented, the caching of the proxy data server could be improved, and so on.
4. Scare students with the prospect of the practical examination or of a timed, closed laboratory session. Independent learning is enhanced if there is the prospect of a closed examination in the future.
5. Hide the book! Naturally, we do not suggest this as a serious strategy. (Even though it may enhance our sales as students scramble to get a copy of a forbidden reference!)

Other Texts

There are many good Java programming texts which are suitable for use with this book and we mention some of them in individual chapters. We have found the texts by Horstmann et al.[15,16,21] and by Hunt[14,32] to be very useful.

Readers cannot do better than to have a personal copy of the seminal work "Design Patterns - Elements of Reusable Object-Oriented Software" by Gamma et al.[28] and there are many up-to-date references on design patterns on the Web and in the literature. See, for example, the patterns home page at `http://hillside.net/patterns/`, as well as the Pattern Languages of Programming (PLoP) workshops. We also provide other references to patterns texts in the body of this book.

Trademarks

Trademarks abound. In computing there are many of them and they disrupt the flow of written text. Readers will have noticed that we have gradually lost the trademark signifier for Java as this Preface has progressed. We will not generally denote trademarks as trademarks in the body of this book. Instead, we list them all here: Java^{TM}, Sun^{TM}, Microsoft^{TM}, Windows^{TM}, Apple^{TM}, Machintosh^{TM}, Cray^{TM}, Vax^{TM}, DEC^{TM}, VMS^{TM}, LabView^{TM}, IDL^{TM}, Globus^{TM}, C#^{TM}.

English Spelling Conventions

After some soul-searching, we have reluctantly decided to use the American-English spelling convention in this book. Both of the authors are of the opinion that the English language would do well to have a universal spelling convention. Computer languages are mostly written by Americans and they are playing a role in entrenching the American view of English spelling around the globe. This book contains an amount of Java code whose keywords and class names follow American English. So it seems logical to adopt this spelling convention for this book. Unfortunately, residual UK English (actually Australian English!) conventions creep into our code examples in method names such as "initialise".

Acknowledgements

Many people have helped us over many years with this project. We particularly wish to thank: Tom Fredian and Josh Stillerman, for being the heart and soul of `MDSplus`, Cesare Taliercio, the coauthor of `jScope`, Giulio Fregonese, for an introduction to design patterns, Rhys Hawkins, Rod Harris, Daniel Kivett, for help developing teaching material, including `EScope`, Boyd Blackwell, for the inspiration to build this case study and for help with `MDSplus` and the `h1data` database, Clive Boughton and Shayne Flint, for an introduction to software engineering and UML in its executable form. Dr Raju Karia is a coauthor of the section of Chapter 20 which describes the porting of `EScope` to a Web server. We also gratefully acknowledge the financial support of the Australian Partnership for Advanced Computing EOT program.

Mark Jarnyk played a major role in teaching courses based on this material including being lecturer-in-charge. He was a patient teacher with a genuine rapport with students. Mark was dedicated to this project and, although ill, made many useful comments on the manuscript. This book is dedicated to his memory.

Canberra, Australia, and Padova, Italy *Henry Gardner*
November 2006 *Gabriele Manduchi*

Contents

Part I Construction of a Waveform Browser

1 e-Science and `EScope` 3
  1.1 What is this Thing Called "e-Science"? 3
  1.2 Computers in Physics 5
  1.3 Computers in Fusion Energy Research 5
  1.4 Programming Languages and Operating Systems 8
  1.5 Our Target Data Server: `MDSplus` 9
  1.6 Two Sample Datasets 10
    1.6.1 The RFX Experiment 10
    1.6.2 The H-1NF Heliac 13
  1.7 jScope and `EScope` 13

2 A Java Client for `MDSplus` 15
  2.1 An Example: `SimplePlot` 15
  2.2 Java IO 17
    2.2.1 A Remark on Exceptions 17
    2.2.2 Character-Based Text Streams 18
    2.2.3 Input from the Keyboard 19
    2.2.4 Writing Text Output 19
    2.2.5 Other Topics in IO 20
  2.3 Exception Handling 20
  2.4 Sockets 21
    2.4.1 A Socket Example: Requesting Data from a Server .. 22
  2.5 Introduction to Threads 25
    2.5.1 Threaded Plot Server 26
  2.6 A Java API for `MDSplus` 28
  2.7 The Data Organization of `MDSplus` 30
    2.7.1 The *mdsip* Protocol for Remote Data Access 32
    2.7.2 Operation of `MDSMessage` 34
    2.7.3 Operation of `MDSNetworkSource` 35

XIV   Contents

    2.8   `PreEScope0`: A Program to Connect to `MDSPlus` 35
    2.9   Programming Exercises 36
    2.10 Further Reading. 37

3 Graphical User Interfaces Using Swing 39
    3.1   Simple GUI Programming 39
         3.1.1 A Blank Frame 39
         3.1.2 Laying-out Components in a `JPanel` 41
         3.1.3 A Bizarre Component Frame 43
    3.2   A Note on Programming Style 45
    3.3   A Look Inside `Plotter`. 46
    3.4   Action Listeners in Swing 47
    3.5   Swing Miscellany 50
         3.5.1 Text Fields and the Model-View-Controller Design
                Pattern 52
    3.6   `PreEScope1`: A Simple GUI for `PreEScope0`. 54
         3.6.1 Using `JOptionPane` to Request Information. 54
    3.7   Programming Exercises 55
    3.8   Further Reading. 56

4 Waveform Graphics. 57
    4.1   Java2D Graphics 57
    4.2   A Second Look at `Plotter` 58
         4.2.1 Basic Set-up 58
         4.2.2 Setting Colors and Strokes. 59
         4.2.3 Transform Data and Plot the Line 60
    4.3   A Fancier Plot 62
         4.3.1 Fonts 63
         4.3.2 Calculating Border Dimensions. 64
         4.3.3 Drawing Titles and Axis Labels 66
         4.3.4 Filtering Data 66
         4.3.5 Overall Structure of `paintComponent` for an Adorned
                Graph 68
    4.4   Axis Calculations: Tick Positions, Tick Values and Scientific
         Notation 69
    4.5   `PreEScope2`: Nicer Graphs from `MDSplus` 71
    4.6   Programming Exercises 71
         4.6.1 Programming Exercises 71
    4.7   Further Reading. 72

5 Interactive Graphics Using Mouse Events. 73
    5.1   Mouse Interfaces and Events 74
         5.1.1 The `MouseListener` Interface 74
         5.1.2 The `MouseMotionListener` Interface. 74
         5.1.3 The `MouseEvent` Class 75

	5.2	PreEscope3: The Graph Point Diagnostic 75
		5.2.1 OpenGL Hardware Acceleration 77
	5.3	Programming Exercises 78
	5.4	Further Reading....................................... 78

6	**Navigating the Database**................................. 79	
	6.1	Custom-Built Linked Lists................................ 80
	6.2	Lists in Swing... 83
		6.2.1 Using the DefaultListModel Class 84
		6.2.2 Using the ListModel Interface...................... 84
		6.2.3 Rendering List Cell Values 85
	6.3	Trees 85
		6.3.1 Recursion 86
		6.3.2 Using Recursion to Probe File Structures............. 87
	6.4	Trees in Swing 88
		6.4.1 Tree Paths and Tree Selection Listeners 88
		6.4.2 Tree Cell Rendering 89
	6.5	MDSTree and MDSTreeNode 89
		6.5.1 Reading the MDSPlus Experiment Hierarchy 91
	6.6	PreEScope4: A Waveform Browser for MDSplus.............. 93
		6.6.1 Issues to Consider for PreEScope4 93
	6.7	Further Reading....................................... 97

Part II Refactoring EScope with Design Patterns

7	**Object-Oriented Analysis and Design** 101	
	7.1	Phases of Software Development............................101
	7.2	UML and Design Patterns 104
	7.3	Design Patterns: Our Approach 105
	7.4	A Diagrammatic Notation: "sUML" 105
		7.4.1 Associations 106
		7.4.2 Association Multiplicities..........................106
		7.4.3 Association Labels................................108
		7.4.4 Reflexive Associations.............................108
		7.4.5 Ignore Aggregation!...............................109
		7.4.6 Dependency 109
		7.4.7 Package Associations 110
		7.4.8 Inheritance and Implementation 110
	7.5	Summary of Our sUML Class Diagrams 111
	7.6	Further Reading.......................................112

8 First Facades 113
- 8.1 Facade 114
- 8.2 EScope0: A "Do Nothing" Code Refactoring Using Packages .. 114
  - 8.2.1 Using Makefiles 118
- 8.3 EScope1: First Implementation of the Facade Pattern 119
  - 8.3.1 Place Facade Interfaces into a Shared Package 119
  - 8.3.2 Facade Interface for the GUI Domain 120
  - 8.3.3 Facade Interface for the Data Server Domain 120
  - 8.3.4 Facade Interface for the Graphics Domain 121
  - 8.3.5 Our Final Product 121
- 8.4 The Mediator Pattern 123
- 8.5 More Notes on Facade and Mediator Patterns 125

9 Adapter 127
- 9.1 Object Adapter Pattern 127
- 9.2 Class Adapter Pattern 129
- 9.3 Are Object Adapters Better Than Class Adapters? 130
- 9.4 EScope2: Sharing Graph Data and Graph Options Between Domains 130
  - 9.4.1 Passing Graph Options from the User Interface 132
  - 9.4.2 An *Articulated Facade* 132
  - 9.4.3 Our Final Product 133
  - 9.4.4 Data Server Domain 133
  - 9.4.5 GUI Domain 134
  - 9.4.6 Graphics Domain 136

10 The Template Pattern 139
- 10.1 Pattern Description 139
- 10.2 EScope3: Splitting up the Graphics Facade 140
  - 10.2.1 The `GraphData` and `GraphMetrics` Classes 143
  - 10.2.2 Drawing Individual Graph Components 143
  - 10.2.3 The Template Pattern for `DrawAxesTicks` 144
  - 10.2.4 Our Final Product 144

11 Decorator 147
- 11.1 Pattern Description 147
- 11.2 EScope4: Adding Zoom and Grab Options Using the Decorator Pattern 149
  - 11.2.1 Grab and Zoom 150
  - 11.2.2 A Mediator Emerges 150
  - 11.2.3 Our Final Product 152

Contents   XVII

12 Patterns at Work: Multiple Waves 155
    12.1 EScope5: Multiple Waveforms 155
        12.1.1 Requirements 155
        12.1.2 Interfaces and External Requirements 156
        12.1.3 Plotting an Array of Waveforms 157
        12.1.4 Modifications to the GUI Domain 158
        12.1.5 Drawing the Cross-Hair 160
    12.2 Our Final Product 162

13 Patterns at Work: Multiple Graphs 165
    13.1 EScope6: Multiple Windows in EScope 165
        13.1.1 Designing for Multiple Windows..................... 167
        13.1.2 The Flexible Grid Layout Manager 169
    13.2 Image Buffering: A Useful Graphics Trick................... 170
    13.3 Our Final Product 171

14 Observer 173
    14.1 Pattern Description...................................... 173
    14.2 EScope7: Integrating Synchronized Interaction in Multiple
        Windows 175
        14.2.1 sharedObserverInterfaces Completes the
            Articulated Facade 175
        14.2.2 Management of a Collection of Graphics Facades 177
        14.2.3 The Graph Scale Interfaces 179
        14.2.4 Our Final Product 180

15 Proxy 181
    15.1 EScope8: Implementation of a Local Data Cache 182
        15.1.1 The DataServerProxy Class........................ 182
        15.1.2 Our Final Product 182

16 State 185
    16.1 Pattern Description...................................... 185
    16.2 Escope9: A State Pattern for the DataServerFacade 186
        16.2.1 Common Interface................................ 187
        16.2.2 State Inner Classes 187
        16.2.3 Managing State Transitions........................ 188
    16.3 Our Final Product....................................... 189

17 Factory Patterns.. 191
    17.1 A Factory Tour 191
        17.1.1 Informal Factory Methods 191
        17.1.2 The Factory Method Pattern 192
        17.1.3 Abstract Factory 192
        17.1.4 Builder 193

		17.1.5 Prototype 193
		17.1.6 Singleton 193
	17.2	EScope10: Multiple Data Servers 193
		17.2.1 Requirements 194
		17.2.2 Implementation 195
		17.2.3 Example Properties File............................ 196
		17.2.4 The `ServerSelectDialog` 197
		17.2.5 The Factory Pattern in EScope10 197
		17.2.6 A Text Data Server 198
	17.3	Our Final Product 199

18 Chain of Responsibility 201
  18.1 EScope11: Avoiding Explicit Connection to Data Servers 201
     18.1.1 An Example Properties File 201
     18.1.2 Implementation 202
     18.1.3 Our Final Product 202

19 Design Patterns and Threads 203
  19.1 Threads and Race Conditions 203
  19.2 Synchronized Methods and `wait()`/`notify()` 206
  19.3 Patterns for Concurrent Systems 208
     19.3.1 The Acceptor-Connector Pattern 209
     19.3.2 The Asynchronous Method Pattern 209
     19.3.3 More Complete Implementations of `AbstractNotifier` 211
     19.3.4 Other Classes in the Asynchronous Method Pattern ... 213
     19.3.5 Summary of the Asynchronous Method Pattern 214
     19.3.6 The Active Object Pattern 214
  19.4 EScope12: A Progress Bar for Downloading Signals 215
     19.4.1 Using Threads with Swing........................... 216
  19.5 Programming Exercises 217
  19.6 Further Reading... 217

20 Postscript 219
  20.1 Design Patterns Then and Now 219
  20.2 The e-Science "Software Stack" 220
  20.3 Server-Side `EScope` for DataGrids (with Raju Karia) 222
     20.3.1 Metadata Indexing, Persistence and Provenance in
            `WebScope` 223
  20.4 A Final Word 225

A Installing and Running Data Servers for `EScope` 227
  A.1 The `MdsipSimulator` Program 227
  A.2 The Text Data Server 228
  A.3 Installing MDSplus 229

　　　　A.3.1　Installing on Microsoft Windows Using the Install
　　　　　　　Shield on the CD 229
　　　　A.3.2　Installing on Windows from www.mdsplus.org 229
　　　　A.3.3　Installing on Linux Using the Supplied RPM 230
　　A.4　Running `MDSplus` with the Sample Data 231
　　A.5　TCL, Traverser and Scope 232
　　　　A.5.1　Creating a Simple Database Using TCL 232
　　　　A.5.2　Examining a Database Using the Traverser 234
　　　　A.5.3　Creating and Viewing Subtrees 235
　　　　A.5.4　Understanding Node Names 236
　　　　A.5.5　Defining Signals and Viewing Them with `jScope` 238
　　　　A.5.6　`UNITS_OF()` and `DIM_OF()` 241

B　Listings of Introductory Examples 243
　　B.1　`BorderComponentFrame` 243
　　B.2　`Plotter` 244
　　B.3　`ShotDataCache2` 246

C　Helper Classes for Accessing `MDSplus` from Java 249
　　C.1　`MDSDescriptor` 249
　　C.2　`MDSDataSource` 251
　　C.3　`MDSNetworkSource` 252
　　C.4　`MDSMessage` 256

D　Listings for PreEScope Examples 263
　　D.1　`PreEScope0` 263
　　D.2　`PreEScope1` 266
　　　　D.2.1　`PreEScope1` Main Program 266
　　　　D.2.2　`EScopeFrame` Class 267
　　　　D.2.3　`Plotter` 271
　　D.3　`PreEScope2` 271
　　D.4　MDSTree and MDSTreeNode 280
　　　　D.4.1　`MDSTreeNode` 280
　　　　D.4.2　`MDSTree` 282
　　D.5　PreEScope4 284
　　　　D.5.1　`EScopeFrame` 285
　　　　D.5.2　`ConnectDialog` 292

E　Listing for EScope4 295
　　E.1　Package Structure 295
　　E.2　Shared Data Interfaces 296
　　E.3　Shared Interfaces 298
　　　　E.3.1　`DataServerFacadeInterface` 298
　　　　E.3.2　`GraphicsFacadeInterface` 299
　　　　E.3.3　`GuiFacadeInterface` 299

	E.3.4 `AbstractGraphicsFacade` 299
E.4	The Data Server Package 300
	E.4.1 `DataServerFacade` 300
	E.4.2 `GraphData` 304
E.5	The Graphics Domain...................................... 305
	E.5.1 The Decorator Classes 306
	E.5.2 Adapter Classes 320
	E.5.3 `GraphMediator` 324
	E.5.4 `GraphicsFacade` 330
E.6	The GUI Domain... 336
	E.6.1 The Dialog Classes 336
	E.6.2 `GraphOptions` 342
	E.6.3 `GuiFacade` 344

F   Excerpts from Later Listings 351
    F.1   EScope5... 351
          F.1.1 `GraphMediator` 351
          F.1.2 `GraphDataInGraphics` 353
    F.2   EScope6... 354
          F.2.1 Reading Properties from `GuiFacade` 354
          F.2.2 `GraphicsFacade` 355
    F.3   EScope7... 360
          F.3.1 `GraphUpdateEvent` 360
          F.3.2 The Scale Interfaces 361
          F.3.3 `GraphicsFacade`: Pop-up Menu and Associated Methods 361
    F.4   EScope8... 363
          F.4.1 `DataServerProxy: getPlotData` 363
    F.5   EScope10.. 364
          F.5.1 `ServerSelectDialog`............................... 364
          F.5.2 `ConnectAction` Inner Class from `GuiFacade` 368
          F.5.3 Factory Interface and Factory Classes 369
          F.5.4 `TextDataServer` 370
    F.6   EScope11.. 375
          F.6.1 `DataServerHandler` 375

References... 379

Index 381

Part I

Construction of a Waveform Browser

1
e-Science and EScope

In this chapter, we will start with a brief history of the new discipline of e-Science. Our goal is to provide some context for the EScope case study which will be developed and refactored using design patterns through the rest of this book. EScope comes from the specific application area of magnetic fusion energy research, so we will beg the reader's indulgence as we describe some of the history of this discipline as well. We will then provide some more background to EScope itself. The main pedagogical treatment of this book does not start until Chapter 2 and it will then follow the structure described in the Preface.

1.1 What is this Thing Called "e-Science"?[1]

"e-Science" is an exciting new buzz-word for "computer science and information technology in the service of science". It is particularly associated with the support of "big" and/or "distributed" science and engineering. It recognizes the revolution in global collaboration which is being wrought by broadband communications and the internet. It also recognizes the steady process of "democratization" which has taken place over some years to the point, now, where a personal computer (PC) in someone's office running a program written on a free operating system might connect to a powerful supercomputer on the other side of the world and a large database of scientific data at yet another remote site. Somewhere in this mix are the scientific instruments which collect the data and which are also, often, connected to the internet and which become part of this global collaboration.

e-Science has rocketed to international prominence as a result of a large program of research and infrastructure development in the United Kingdom [1]. The vision of this program emphasized the development of a "Grid" of

---

[1] With apologies to "What Is This Thing Called Science?" – a lovely book on the philosophy of science by Alan F. Chalmers.

computational resources which could be coordinated and made available to scientists and engineers. This notion of a Grid has been strongly associated with research in the USA which has resulted in the development of a basic grid-computing toolkit, called Globus [2]. These days, there are conferences on e-Science and many of the papers are on building scientific applications for distributed, heterogeneous, computing and data resources (such as the Grid), on the scheduling and management of tasks on the Grid, on the management of Grid-like computing resources themselves or on the development of distributed, networked data stores known as "DataGrids". These contemporary concerns of e-Science will be described in Chapter 20.

But, at the end of the day, e-Scientists still need to program. The programs that they write will still involve elements of numerical computation, visualization and data-processing. If they are to program effectively in the new environment of a distributed Grid, it would be handy for them to have some experience in a modern programming language which includes visualization and networking as part of its basic fabric. A good choice is Java, and it has been chosen for the case study in this book.

In 2001 a teaching program in "eScience" [3] was started at the Australian National University (ANU) in Canberra (under the direction of one of the authors of this book). This program was oriented towards students with a prior degree in science or engineering and aimed to provide them with a mix of courses which defined an interpretation of e-Science:

- Algorithms for High Performance Computing optimization
- Visualization and computer graphics
- Networking
- Graphical, statistical data analysis
- Programming and software engineering

The ANU e-Science program endeavored to have a strong emphasis on software engineering and good programming practice. There is a body of opinion which is of the view that writing scientific software is hard and that it lags well behind best commercial practice.[2] After some years of contributing to the e-Science education program at ANU, we are convinced that we have made progress towards remedying this bleak situation. One of main principles is to apply and adapt some exciting developments in computer science, known as "design patterns", to the design and construction of e-Science software. This is the rationale for this book.

---

[2] See, for example, the report of the US President's Information Technology Advisory Committee, "Computational Science: Ensuring America's Competitiveness", June 2005 [4].

1.2 Computers in Physics

From their earliest days, computers have been been used to support theoretical and experimental research in physics and related areas. Although the power of the first computers was laughable compared with today's machines, they ushered in a new epoch in the methods used to investigate and simulate phenomena. In turn, these new simulation techniques have fed back into the evolution of computer software and hardware. For example, the Monte Carlo method was developed by physicists and mathematicians working with the earliest electronic digital computers in the 1940s. Monte Carlo methods are now used in fields as diverse as particle physics, financial modelling, nuclear reactor design and computer games!

In the early days, numerical analysis and simulation, usually carried out in batch sessions, was the main mode of application of computers to scientific research. Later, computers were used to provide real-time support for experiments and, by the 1970s, computer-based data-acquisition systems started to become commonplace. Experimental results which had previously been displayed as oscilloscope traces were able to be directly transferred from sensor equipment into computer memory. During this period, the "Computer Aided Measurement And Control", or CAMAC, standard (also known as IEEE 583) was developed to specify common strategies for connecting measurement and control devices to computers. It is still widely used today.

Since the 1990s, thanks to their continuous improvement in speed and memory capacity, computers have started to be used to provide *active control* of physical phenomena during experiments. This has realized new scenarios in scientific research where computers can, in real time, provide support to experiments by acquiring sensor data, applying sophisticated control algorithms, and generating a set of output signals for the control circuitry.

Also since the 1990s, World Wide Web technology has been widely employed in physics research. As well as supporting the familiar activities of scientific scholarship, the Web is increasingly used for the development of distributed user interfaces in large experiments and for exchanging data and results among laboratories. In fact, physicists are proud to note that it was a project at the European particle physics laboratory, CERN, in 1989, which gave rise to the World Wide Web itself!

1.3 Computers in Fusion Energy Research

Magnetic fusion research is a field of science and engineering which aims to develop a future source of energy for the large-scale generation of electricity.

In a nuclear fusion reactor, a gas formed of heavy isotopes of hydrogen, such as deuterium and tritium, will be heated to temperatures above 100 million degrees Centigrade. Under these conditions, the kinetic energy of the positively charged nuclei will be sufficient to overcome their Coulomb repulsion so that many nuclei will "fuse together". In the case of deuterium and tritium, this fusion will produce a helium nucleus, a neutron, and more energy. The fusion of hydrogen isotopes occurs naturally in the sun and is responsible for the production the energy that sustains all life on earth.

The advantages of producing electricity by means of controlled nuclear fusion include the following:

- Nuclear fusion reactions do not produce carbon dioxide or any other greenhouse gas.
- Unlike nuclear fission, nuclear fusion is intrinsically safe. If something were to go wrong in a fusion reactor, then the nuclear reaction spontaneously terminates. Disasters comparable to a meltdown of the core of a nuclear fission reactor can never occur.
- The problems of disposal of nuclear waste from fission power plants would be significantly reduced with nuclear fusion technology. This is because fusion reactions do not, themselves, produce long-lived radioactive products.
- Nuclear fusion reactors do not, themselves, produce material which can be used for the construction of nuclear weapons.
- The raw fuel for fusion reactors is relatively cheap and uniformly distributed around the world.

The major problem with getting controlled nuclear fusion to work is the confinement of the gas at the very high temperatures needed. Because, at these temperatures, the gas will be a fully ionized *plasma* of (negatively-charged) electrons and (positively-charged) nuclei, strong magnetic fields can be used to guide the orbits of these charged particles. This kind of confinement is called magnetic confinement and it currently represents the major line of fusion research.

The idea of harnessing the power of nuclear fusion for the peaceful generation of electricity dates from the 1950s. In spite of being the butt of endless "only 20 years away" jokes, the field is one of enthralling progress which has revealed fundamental understanding of cooperative phenomena as well as important engineering advances. Even during the height of the cold war, magnetic fusion research has been a field of great international scientific cooperation. Today there are several experimental, magnetic-fusion devices around the world and most of these are toroidal ("doughnut-like") magnetic geometries. At the time of writing, the largest of these is a "tokamak" called the Joint European Torus which is located in the United Kingdom [5]. In 2005, an international consortium of countries [6] (the European Community, USA, Russia, Japan, China and Korea) agreed to construct the first prototype fusion reactor, called "ITER" at Cadarache, France, at a cost of some US$10 billion. As shown in Fig. 1.1, this will be a massive device! (The small figure at the bottom is a representative human!)

So, fusion experiments are "big science" and they make extensive use of computers. Numerical simulation codes are used for the physics design of new

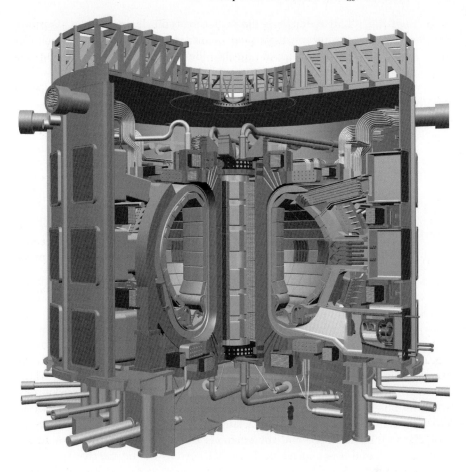

Fig. 1.1. Cut-away diagram of the ITER tokamak. Published with permission of ITER.

magnetic configurations, for the engineering design of the magnetic coils and support structures and for the interpretation of experimental results. Data-acquisition systems supervise the set-up and control of experiments and manage the acquisition of data by hundreds of diagnostics. Database management systems are used to handle the large amount of raw, experimental data and the processed data produced by simulation programs.

In present-day, magnetic-fusion experiments, plasma is confined for times ranging from hundreds of milliseconds to several seconds. During this time, which is called a *"shot"* (or a *"pulse"*), data is collected and stored in local memory by the data-acquisition devices. After the shot, the data is transferred to a *"shot file"*. This specialized file structure typically stores signals representing the time-evolution of physical quantities on time-scales as small as microseconds. The typical size of one shot file can be hundreds of megabytes and, during the lifetime of an experiment, many terabytes of shot files will be accumulated. For the large, international experiments there is a critical need to have this data accessible, and usable, by scientific collaborators around the world in a timely fashion. It is an example of an e-Science DataGrid.

1.4 Programming Languages and Operating Systems

The evolution of software technology in fusion energy research is a mirror of the general trends in big-infrastructure science and engineering disciplines. The points we make in this section are generally true for areas such as astronomy, space science, aeronautics, particle physics, and so on.

Before the 1970s, programs written in FORTRAN, and executed in a batch mode, represented the large majority of the computer use in fusion research. The FORTRAN language was also used for the first, large, data-acquisition systems based on the Digital Equipment Corporation "PDP", and later "VAX", architectures. Outside of "Cray"-based supercomputing centers, the operating system of VAX computers, "VMS", was probably the major platform in the 1980s. The C programming language started to be widely used for the development of data-acquisition software from the second half of the 1980s while FORTRAN (and its modern variants such as FORTRAN90/95) has continued to be important for numerical simulation and data analysis. Object-oriented (OO) programming languages, mostly C++, appeared in fusion research in the 1990s and the language zoo has now expanded to include Java, Python and other specialized systems such as the Interactive Data Language [7] and LabView [8].

Even though computer hardware and software evolve continuously, the large investment required for a control and data-acquisition system for big-infrastructure science experiments, has meant that their system architectures have normally been frozen during the lifetime of the project (often more than 10 years). However, several compelling reasons to modify these systems have emerged in recent years:

- It has turned out that the use of proprietary hardware platforms and operating systems, which was the rule until the 1990s, has forced users to become tied to solutions which are non-optimal or, even worse, no longer supported. This has been the case for operating systems as well as for application software. In response, there has been a rapid growth in the use of Linux platforms in fusion research. The fact that Linux is free and open source has meant that systems can be made adaptable to specific scientific requirements. As an example, real-time extensions of the Linux kernel are now often used in digital control systems.

- Interoperability among different computer platforms is now becoming an important factor in research. In a typical scenario, which is actually implemented on the data-acquisition system of the RFX experiment described in Section 1.6.1, Linux PCs are used for data acquisition and storage and for real-time control, while a mixture of Linux and Microsoft Windows PCs are used in the control room for hosting user interfaces. Individual scientists then use a mixture of Linux, Windows and Apple Macintosh personal computers for remote data analysis.
- As experiments become larger and more internationalized, there is a growing imperative for web-based, remote access to data acquisition and control systems and the e-Science DataGrids based on them.

All of these considerations motivate the adoption of Java for part of the software fabric of a large science experiment. To take the issue of interoperability as an example, two important aspects of complicated software systems are quite difficult to port across operating system platforms: multi-threading and graphical user interfaces. In both cases, the Java Virtual Machine model offers improvements over direct calls to native libraries. In the case of the data-acquisition and control system for the fusion experiment RFX (described in Section 1.6.1) one Java program delivers the synchronization and scheduling of an entire local network of different computers and another Java program provides a waveform-browser interface for the control room and for remote data analysis. Over almost 10 years of development of the RFX system, the experience has been that the boundary between the, Java-based, front-end software systems and the, C-based, back-end systems has been pushed progressively back. That is, as time has gone by more and more of the entire system has been able to adopt interoperable, Java components.

1.5 Our Target Data Server: MDSplus

An open-source software project specific to fusion research is the MDSplus ("Model Data System") data-acquisition system [9]. The major contributors to MDSplus have been members of the following scientific laboratories:
- MIT Plasma Science and Fusion Center, Cambridge,MA, USA[3]
- Los Alamos National Laboratory, Los Alamos, NM, USA
- Istituto Gas Ionizzati, Padua, Italy

The MDSplus system, originally developed in C for the VAX/VMS operating platform, has been in use since 1990 and it represents the first shared-software facility for control and data acquisition in the fusion research community. Large parts of the system have been since been ported to several flavors of UNIX as well as to Windows. It is currently used in many magnetic fusion experiments around the world. Its data access layer, mdsip, which defines

---

[3] The main authors of MDSplus are Tom Fredian and Josh Stillerman.

the database organization, the data types and the network protocol for distributed data access, has become a "de facto" standard for the exchange of experimental data among fusion laboratories.

The main components of MDSplus are

- A database management system for storing and retrieving data in tree-structured shot files.
- An integrated environment for handling a variety of data types and for processing data on-the-fly. `MDSplus` anticipated that data-acquisition program modules would need to deal with very different types of data, ranging from simple scalars to signals describing the evolution of some quantity over time, and even to video images and operator comments. `MDSplus` enables different data types to be combined to form complex expressions. For example, a raw signal can be scaled and offset and combined with another signal before being delivered to a client program.
- A data-acquisition engine which uses information stored in the pulse database to supervise the coordinated execution of control, data acquisition and online analysis of a distributed software system.
- A network layer for exporting the content of the local shot databases so that they are available over the internet.
- Graphical user interfaces for the configuration of the experiment parameters and for the display of acquired results. Results stored as signals are displayed as oscilloscope-like waveform.
- A version of `MDSplus` is Globus-enabled for execution on the Grid.

MDSplus is freely available from its home web-site ([9]). Appendix A describes how you can go about installing it on a personal computer.

1.6 Two Sample Datasets

The real-life datasets used in the examples in this book are taken from two fusion experiments which we, the authors, have worked on. Both are "medium sized" toroidal magnetic geometries but neither is a tokamak like ITER. Once again, we beg the reader's indulgence as we take some space to describe these experiments and the datasets that we have taken from them.

1.6.1 The RFX Experiment

The RFX "reverse field pinch" [10] is a magnetic fusion experiment located in Padua, Italy. Its magnetic configuration differs from tokamaks in that the very large currents flowing through the plasma mean that equilibrium operation will not be possible without active control. So, one of the major characteristics of RFX is the extensive use of real-time feedback and control systems.

After a closure for 5 years for a significant upgrade, the RFX experiment recommenced operation in 2005. Its results will become part of the scientific database of tokamaks and other alternative fusion concepts and will contribute to the design and operation of ITER and subsequent fusion reactors.

MDSplus is tightly integrated into the operation of RFX. For example, a view of the RFX operator interface is shown in Fig. 1.2, a view of the waveform browser is shown in Fig. 1.3 and the control room, itself, is shown in Fig. 1.4.

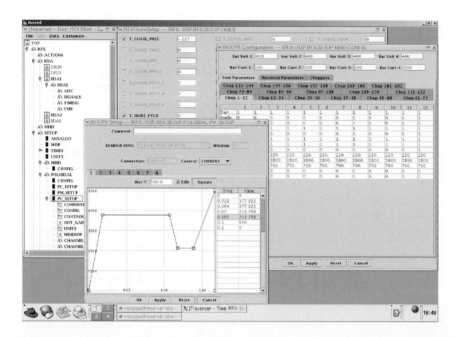

Fig. 1.2. One of the screens from the RFX operator interface. The tree displayed on the left allows navigation into the configuration structure of the experiment which is composed of thousands of items.

In a typical RFX scenario, operational staff need to set-up the configuration for the current experimental campaign. Hundreds of parameters are usually involved in the management of any experimental configuration. Good, graphical user interfaces are, therefore, important for providing an easy configuration definition and and consistency checks. Good user interfaces are also important for the fast and effective visualization of experimental results, so that scientists in the control room can react quickly to determine how to proceed with the next shot. These user interfaces for the RFX data-acquisition and control systems are all part of the MDSplus family of software.

Part of an RFX shot file is supplied on the accompanying CD for use with the EScope case study. This file contains data and set-up parameters for the array of "poloidal" and "toroidal" field coils for shot 17615 which took place in late 2005. This RFX file tree is labeled "EDAM" for "Engineering Data Acquisition Machine".

12   1 e-Science and EScope

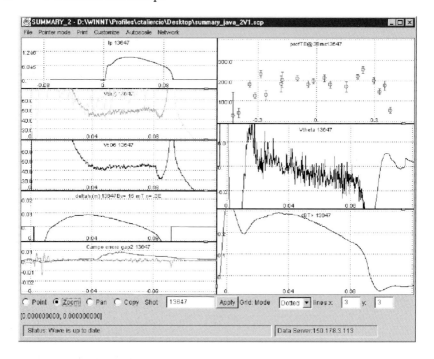

Fig. 1.3. A sample jScope interface.

Fig. 1.4. Using jScope in the RFX control room.

1.6.2 The H-1NF Heliac

The H-1NF heliac [11] is a "stellarator" fusion experiment located at the Australian National University. Although its magnetic configuration is still toroidal, like a tokamak or reverse field pinch, its structure is much more three-dimensional. The idea is that the "twist" needed to produce a confining magnetic field can be created by the configuration itself rather than driving large currents through the plasma.

Part of an H-1NF shot database, for shot 37025, is also included on the accompanying CD for use with EScope. The signals of interest are to do with the "operations" information for this shot (the current in the external coils, the plasma diamagnetic signal and so on) and for the radio-frequency heating. Many of the programming examples in this book refer to this shot file.

1.7 jScope and EScope

One of the user-interface components of MDSplus is a tool written in Java called jScope. jScope takes full advantage of the platform independence of the Java Virtual Machine and provides interactive graphical display of waveforms. Once waveforms have been displayed, users can manipulate them by, for example, zooming regions of interest or moving a cross-hair over selected waveforms to retrieve the exact X and Y data values.

jScope is normally used during the experimental sessions of RFX to provide a quick and effective visualization of what is happening during the experiment. A sample interface is shown in Fig. 1.3.

jScope can run across different platforms and it can be connected to a a variety of data sources, Another feature of jScope is the possibility of integrating new kind of data sources by implementing plug-ins. Several kinds of data-source plug-ins are currently available and this makes it possible to look at data coming from many different fusion experiments around the world.

Figure 1.4 shows jScope in use in the control room of the RFX experiment, where jScope clients run both on Linux and Windows PCs and retrieve experimental data from a Linux server which hosts the shot files.

jScope is still evolving to integrate new graphical functionality as well as adding new data sources. The current implementation is about 30 thousand lines of Java code, and it would be too complicated to be presented as a case study. For this reason, a more simplified tool, EScope, will be developed throughout this book. EScope provides a graphical interface which is very similar to jScope and which also allows the integration of different data sources. It illustrates many of the most interesting aspects of jScope, but it is simple enough to be constructed, refactored, extended (and understood!) by students.

2

A Java Client for MDSplus

This book assumes that you, our readers, have had an introductory-level experience with Java programming. You should have written some Java classes and methods and you should have built up a small, multi-class system. The next few sections of this chapter are intended as a revision of the parts of the Java language, and API, that you need in order to write client/server programs: specifically Java's input/output framework including sockets. Following this, we will briefly discuss exception handling and threading and then we will start to build a program which can interact with an MDSplus server. This program will build on itself to eventually become our prototype EScope application. We will call its various versions "PreEScope" to distinguish them from the EScope versions which we will develop through the successive application of design patterns.

The Java input/output (IO) framework makes it possible to get data from a server over a network in a completely analogous way to reading it from a file on your local disk. So we will start out by looking at Java IO in general and we will do this using an example.

2.1 An Example: SimplePlot

The code in Listing 2.1 reads in two arrays, of X and Y data, and then calls a static method of a Plotter class to plot the graph. This example and its data files are available on the CD provided with this book (as are all of the other examples). You can compile and run it providing its class file is located in the same directory as the data files.

For the moment, we are not concerned with the details of the plotting method or about the details of the data itself. We just observe that this program opens up two files called "data_xVals" and "data_yVals". These files contain *binary* data; you cannot list them at a terminal but you can read them into your Java program using classes from the java.io package. The files each contain an int (giving the length of the data array) followed by an array of double data. The example foreshadows the way that EScope will eventually work!

Listing 2.1. SimplePlot reads binary data and constructs an XY plot.

```java
import java.io.*;
/** Read data from 2 binary files and call Plotter.plot*/
public class SimplePlot
{
    public static void main(String[] args)
    {
        double[] xVals=null, yVals=null;
        int xLen=0, yLen=0;
        try
        {   // read Y values from first file
            File f1 = new File("data_yVals");
            DataInputStream in1 =
                new DataInputStream(
                new BufferedInputStream(
                new FileInputStream(f1)));
            yLen = in1.readInt();
            yVals=new double[yLen];
            for (int i = 0; i < yLen; i++)
            {
                yVals[i] = in1.readDouble();
            }
            // read X values from second file
            File f2 = new File("data_xVals");
            DataInputStream in2 =
                new DataInputStream(
                new BufferedInputStream(
                new FileInputStream(f2)));
            xLen = in2.readInt();
            xVals=new double[xLen];
            for (int i = 0; i < xLen; i++)
            {
                xVals[i] = in2.readDouble();
            }
            in1.close();
            in2.close();
        }
        catch (IOException e) {System.out.println(e);}
        Plotter.plot(xVals, yVals, "test");
    }
}
```

2.2 Java IO

Java "streams" are objects which deal with the flows of data into, and out of, your program:

- "Input streams" get sequences of bytes from a "source" of data.
- "Output streams" send sequences of bytes to a "sink" of data.

Streams can transfer data to and from files, network connections and other input/output devices as well as internal program variables.
To read data from a *file* we need to

- open the file as a `FileInputStream`
- apply some buffering to our stream using, for example, objects of type `BufferedInputStream`
- call methods of a special "reader" object which can recognize specific data types (for example, the `DataInputStream` methods `readInt()` and `readDouble()` read and recognize `int` and `double` data respectively)

The complete "reader object" is constructed by "chaining together" the opening, buffering and reading objects. The constructor of the reader object accepts a buffering object which itself accepts an opening object in its constructor. For example, in the following code excerpt from `SimplePlot`, `FileInputStream` reads bytes from the file. `BufferedInputStream` buffers the data so that not every read is associated with a disk access and `DataInputStream` reads groups of bytes that are associated with the primitive data types:

```
File f2 = new File("data_yVals");
DataInputStream in2 =
    new DataInputStream(
    new BufferedInputStream(
    new FileInputStream(f2)));
```

Note that binary files can combine data of different types. In the example, we read an `int` which let us know how many elements of a `double` array remained to be read.

In addition to `readInt()` and `readDouble()`, `DataInputStream` has analogous methods to read all the primitive types as well as `readUTF()` which returns a string. It can also read whole lines, read whole files and skip over bytes of data.

The unusual pattern of constructing a `DataInputStream` object by feeding it objects of the other classes is, in fact, a well-known design pattern known as the *decorator*. We will return to it much later in this book in Chapter 11. A schematic representation of the chaining involved in the pattern is shown in Fig. 2.1.

2.2.1 A Remark on Exceptions

All of the file handling methods described above throw *exceptions* when unforeseen error conditions occur. For example, your code may attempt to read

Fig. 2.1. Schematic representation of a `DataInputStream` object showing the chaining to objects of type `BufferedInputStream` and `FileInputStream`.

data from a file which does not exist. The block of code which contains the read method will stop executing and control will transfer to a appropriate `catch` block. Java forces you to wrap methods which throw exceptions inside `try...catch` blocks like the ones shown in the `SimplePlot` example. We will discuss exception handling in Section 2.3.

2.2.2 Character-Based Text Streams

Byte streams are efficient ways of storing data but they are not readable by humans! We can use text files to cope with this problem.

Using text files with Java programs creates another problem. Java uses Unicode (2 byte) encoding which is not the usual operating-system default. Unicode is designed to be compatible with all of the human-readable scripts in the world so it contains additional storage requirements to the common, ASCII, encoding.

A Java `InputStreamReader` object turns an input stream that is based on an operating-system default into one that it based on Unicode. Like the `BufferedInputStream`, it is constructed using the Decorator pattern. An argument of type `FileInputStream` is passed to the constructor:

```
InputStreamReader in =
    new InputStreamReader(new FileInputStream("textData"));
```

Once you have created your `InputStreamReader` you can pass it to the constructor of an object of the `BufferedReader` class which can read whole lines of text and return them as strings:

```
BufferedReader in1 = new BufferedReader(
                     new InputStreamReader(
                     new FileInputStream("textData")));
String line;
while ((line=in1.readLine()) != null)
{
    ....
}
```

Now that you have your data as a string, you might need to parse the strings to convert them to other data types. For example,

```
double x = Double.parseDouble(line)
```

would be a valid way of parsing a line which contains just one **double** value. Otherwise, if there are several numbers on any one line you could use a `java.util.StringTokenizer` object to break up the line and then parse each token separately.

2.2.3 Input from the Keyboard

As something of an aside, we remark on the way you *used to have to* read input from your keyboard into a Java program. Although the procedure is somewhat tedious and now unnecessary, it is interesting because of the neat way that it fits in with the other Java IO examples shown here.

Before Java 1.5, you needed to be aware that the `System` class contains a static field, `System.in`, which is an object of type `InputStream`. It also has a static field, `System.out`, which is an object of type `PrintStream`. `InputStream` objects have a method, `read`, which is able to read a single byte at a time but it cannot read a line of text input (which is what you want to do). In order to increase the functionality of `System.in`, it needs to be passed to an `InputStreamReader` object which can read individual characters. This `InputStreamReader` object then needs to be passed to a `BufferedReader` object which has a `readLine` method to read entire lines at once.

This is how it works:

```
InputStreamReader isReader = new InputStreamReader
                                            (System.in);
BufferedReader bReader = new BufferedReader(isReader);
System.out.println(''What is your name?'');
String name = bReader.readLine();
```

Since Java 1.5, this has been greatly streamlined. There is a class `java.util.Scanner` which can be used to deliver string tokens corresponding to successive lines of input typed at a console and this is how it works:

```
java.util.Scanner sc = new java.util.Scanner(System.in);
System.out.println(''What is your name?'');
String name = sc.next();
```

2.2.4 Writing Text Output

The `PrintWriter` class is often used for writing text to files:

```
PrintWriter out = new PrintWriter(
                    new OutputStreamWriter(
                    new FileOutputStream("output.txt")));
```

A `PrintWriter` object has `print` and `println` methods (just like `System.out`). It also has `flush()` and `close()` methods which are good to use to make sure that the output buffers have been emptied. The `PrintWriter` constructor can also be called with a second boolean argument set to **true** in order to "autoflush" buffers.

Note that you can actually leave out the `OutputStreamWriter` because `PrintWriter` adds one by default. (But, in contrast, `BufferedReader` must have `InputStreamReader` in order to read in text data.) The three classes are chained together using the decorator pattern just the same way as for the `InputStream` example of Fig. 2.1.

2.2.5 Other Topics in IO

Java's Input/Output framework has many other subtleties and features which you can explore using Sun's online Java tutorials and documentation and by looking at good books. Some interesting ones are

- parsing strings
- reading from zip and jar compressed files
- reading from URL's

2.3 Exception Handling

The sorts of exceptions that a programmer needs to handle using Java's exception mechanism include

- user input errors,
- trying to read past the end of a file,
- trying to open a file which does not exist,
- trying to send data to a socket which has not been opened.

All of these problems occur when the behavior of the system, or a user, causes an error condition to occur. These errors are quite different from programming errors. As a pedagogical point, you should fix your bugs before your program gets released. You should *not use exceptions* to deal with the existence of possible programming bugs!

Exceptions are objects of classes which extend the `Throwable` class. You can define your own exception classes. Java has an extensive hierarchy of exception classes which you can read about as you need them.

Exceptions are *thrown* by methods and *caught* within the code bodies of other methods: If method B throws an exception and it gets called by another method, A, then method A must either

- throw the same exception as B, or
- the call to method B must take place inside a `try` ... `catch` block inside method A.

In the `SimplePlot` example, all of the input-output calls are placed inside one big `try` block in the calling method. Because all of the exceptions of the `java.io` library are children of `IOException`, they end up being handled by the one `catch` clause:

```
catch (IOException e) {System.out.println(e);}
```

This `catch` clause simply printed out a string identifier of the offending exception object. Using the parent `IOException` object is a very "coarse-grained" way of catching IO exceptions. As an alternative, it is possible to have a number of `catch` clauses following one `try` block. Any number of exception subclasses of `IOException` could be identified one after another and the program would be able to test each one and provide specialized handling for each.

The Java exception-handling mechanism allows you to look at only the `try` blocks to identify which parts of a method get executed if all goes well. This is a great advantage over other languages where you need to "pollute" your algorithms with "what if something bad happens here" types of statements in order to deal with user or system errors.

If you want to *throw* an exception from your method, you add a `throws` clause to the header such as

```
public void handleInput () throws IOException
```

You can keep throwing exceptions right up to a program's `main` method. If the `main` method throws an exception, then it gets handled by the Java virtual machine.

There is also an optional `finally` clause which follows the `catch` clauses in a method which handles exceptions. It contains a code block which gets executed after all of the `try`, and possibly `catch`, clauses have been processed (i.e. regardless of whether exceptions were generated during the processing of the `try` block).

2.4 Sockets

A *process* is an active computer program on some computer somewhere in the world. We are all familiar with personal computers which run many different processes (your program, the desktop, a network connection, email...) at the same time. Before the widespread use of parallel supercomputers, scientific programming was concerned with single-process applications which solved problems using a well-defined sequence of steps. On the other hand, much real-time programming for data acquisition and experimental control has needed to deal with *concurrent processes* which are active at the same time. Internet programming has some similarities with real-time systems where the concurrent processes can be located on distant computers.

Concurrent processes on the same or different computers can communicate using "sockets". The Java model for sockets treats them in a very similar way to files. Each socket can be associated with an input stream and an output stream; when your program sends data to a socket's output stream it looks just like you are writing to a file. But, in reality, your data might be being shipped all over the world. (The Java API also contains special classes for communicating with web pages and you can even configure a Java program to send people email.)

In order to establish a socket for communication to a remote server, a Java program needs to know the Internet Protocol (IP) address, and the port number, on the remote computer. Once your process (the "client") has connected to the server then both processes are, in some sense, locked together.

2.4.1 A Socket Example: Requesting Data from a Server

In the following listing `SimplePlotClient` sends requests for data over the internet and then plots the data as before. `SimplePlotClient` uses the IP "localhost" address of the present computer and the port number 8004 (port numbers in the 8000's are good ones to use for your networking experiments). The requests are sent using a very simple "language" having only 3 commands. Note that the server can accept these commands in any order.

YVALS Asks the server to return an `int` which gives the number of Y values. This is followed by the array of y values.
XVALS Requests the array of X values in the same way as for the Y values.
CLOSE Closes the connection.

Listing 2.2. `SimplePlotClient` reads binary data from a socket and plots it.

```
import java.io.*;
import java.net.*;
/* Interacts with the simple plot server */
public class SimplePlotClient
{
    public static void main(String [] args)
    {
        double [] xVals=null, yVals=null;
        int xLen=0, yLen=0;

        try
        {
            Socket s = new Socket("localhost",
                                    8004);
            PrintWriter out =
                new PrintWriter(s.getOutputStream(), true);

            DataInputStream in = new DataInputStream(
                                new BufferedInputStream(
                                    s.getInputStream()));
            out.println("YVALS");
            yLen = in.readInt();
            System.out.println("yLen = " + yLen);
            yVals=new double[yLen];
            for (int i = 0; i < yLen; i ++ )
```

```java
        {
            yVals[i] = in.readDouble();
            if (i%100 == 0) System.out.println(yVals[i]);
        }
        out.println("XVALS");
        xLen = in.readInt();
        System.out.println("xLen = " + xLen);
        xVals=new double[xLen];
        for (int i = 0; i < xLen; i ++ )
        {
            xVals[i] = in.readDouble();
            if (i%100 == 0) System.out.println(xVals[i]);
        }
        out.println("CLOSE");
        s.close();
        Plotter.plot(xVals,yVals,"test");
    }
    catch (IOException e) {System.out.println(e);}
    }
}
```

Here is the corresponding server program to Listing 2.2. Look at the two programs carefully and see how they correspond:

Listing 2.3. `SimplePlotServer` is the corresponding server class to `SimplePlotClient`.

```java
import java.io.*;
import java.net.*;
public class SimplePlotServer
{
    public static void main(String[] args)
    {
        SimplePlotServer s = new SimplePlotServer();
    }
    public SimplePlotServer()
    {
        double[] xVals=null, yVals=null;
        int xLen=0,yLen=0;
        try
        {
            ServerSocket s = new ServerSocket(8004);
            Socket connection = s.accept();

            BufferedReader in = new BufferedReader
                (new InputStreamReader(
                           connection.getInputStream()));
            DataOutputStream out =
                       new DataOutputStream(
                       new BufferedOutputStream(
```

```java
                    connection.getOutputStream()));
boolean done=false;
while (!done)
{
    String line = in.readLine();
    if (line==null) done=true;
    else
    {
        if (line.trim().equals("YVALS"))
        {   // Read data from file
            File f2 = new File("data_yVals");
            DataInputStream in2 =
                new DataInputStream(
                new BufferedInputStream(
                new FileInputStream(f2)));
            yLen = in2.readInt();
            System.out.println("yLen = " + yLen);
            yVals=new double[yLen];
            for (int i = 0; i < yLen; i++)
            {
                yVals[i] = in2.readDouble();
            }
            // Write data down the socket
            out.writeInt(yLen);
            for (int i = 0; i < yLen; i++)
            {
                out.writeDouble(yVals[i]);
            }
            out.flush();
            in2.close();
        }
        if (line.trim().equals("XVALS"))
        {   // Read data from file
            File f3 = new File("data_xVals");
            DataInputStream in3 =
                new DataInputStream(
                new BufferedInputStream(
                new FileInputStream(f3)));
            xLen = in3.readInt();
            System.out.println("xLen = " + xLen);
            xVals=new double[xLen];
            for (int i = 0; i < xLen; i++)
            {
                xVals[i] = in3.readDouble();
            }
            // Write data down the socket
            out.writeInt(xLen);
            for (int i = 0; i < xLen; i++)
            {
```

```
                        out.writeDouble(xVals[i]);
                    }
                    out.flush();
                    in3.close();
                }
                if (line.trim().equals("CLOSE"))
                                            done = true;
                }
            }
            out.flush();
            connection.close();
        }
        catch (IOException e)
        {
            System.out.println(
                    "Error in Simple Plot Server " + e);
        }
        System.exit(0);
    }
}
```

2.5 Introduction to Threads

We will briefly consider the subject of threads in Java. This is so that you can have a better idea of what the MDSPlus server is doing when it interacts with multiple clients at the same time. The subject of threads in Java is important and it can be quite complicated. This brief excursion into its territory might be skipped over on a first reading of this book. Multi-threaded programs will be the subject of Chapter 19.

Threads are like independent processes except that threads share the same data context (memory space) of the program which started them. Threads are easy to create and destroy and communication between threads is much easier than between processes. They are, therefore, useful candidates for parallel processing.

There are two ways to use threads in Java:

- to extend the Thread class, or
- to implement the Runnable interface.

The second method is available to you if the class you want to make into a thread already extends another class.

To start a new thread, you can create an instance of a class which extends Thread or implements Runnable. This class *must have* a run method. You then start the run method with a call to start()! This very strange pattern is because a distinction needs to be made between starting an *independent thread* and making a direct call to **run()** which would start it in the *same thread* as the caller.

2.5.1 Threaded Plot Server

Recall that our SimplePlotServer program of Sect. 2.4.1 had the following lines:

```
ServerSocket s = new ServerSocket(8004);
Socket connection = s.accept();
```

The server waits for a client to connect at the line **s.accept**. Once a client has connected, input and output streams are allocated and the server listens for a set of commands. In order to convert this program into a threaded plot server, we can encapsulate the server logic in the **run()** method of a new class which extends **Thread** and we spawn threads as follows:

```
try
{
    ServerSocket s = new ServerSocket(8004);
    do
    {
        Socket connection = s.accept();
        ThreadedPlotDataReader reader =
                new ThreadedPlotDataReader(connection);
        reader.start();
    }
    while (true);
}
```

This server program now functions in an infinite loop and you will need to crash it if you want it to stop! (Try typing Crtl-C.)

The **SimplePlotThreadedServer** class is shown below. You can test it out by running several clients at the same time in several different shell windows (remembering to crash the server when you have finished!) The **mdsip** simulator program, described in Appendix A.1, on page 227, has a structure which is very similar to SimplePlotThreadedServer.

Listing 2.4. Threaded version of a plot server.

```
import java.io.*;
import java.net.*;
public class SimplePlotThreadedServer
{
    public static void main(String[] args)
    {
        SimplePlotThreadedServer s =
                        new SimplePlotThreadedServer();
    }
    public SimplePlotThreadedServer()
```

```java
    {
        try
        {
            ServerSocket s = new ServerSocket(8004);
            do
            {
                Socket connection = s.accept();
                ThreadedPlotDataReader reader =
                    new ThreadedPlotDataReader(connection);
                reader.start();
            }
            while (true);
        }
        catch (IOException e)
        {
            System.out.println
                ("Error in SimplePlotThreadedServer " + e);
        }
        System.exit(0);
    }
}
class ThreadedPlotDataReader extends Thread
{
    Socket connection;
    double[] xVals=null, yVals=null;
    int xLen=0, yLen=0;
    public ThreadedPlotDataReader(Socket s)
    {
        this.connection = s;
    }
    public void run()
    {
        try
        {
            BufferedReader in = new BufferedReader(
                new InputStreamReader(
                connection.getInputStream()));

            DataOutputStream out =
                        new DataOutputStream(
                        new BufferedOutputStream(
                        connection.getOutputStream()));
            boolean done=false;
            while (!done)
            {
                String line = in.readLine();
                if (line==null) done=true;
                else
                {
```

```
                    if ( line . trim (). equals ("YVALS" ))
                .....  // (same as SimplePlotServer)
                    if ( line . trim (). equals (" CLOSE" ))
                                                done = true;
                }
            }
            out . flush ();
            connection . close ();
        }
        catch ( IOException e)
        {
            System . out . println (
                    "IO error in ThreadedPlotDataReader" );
        }
    }
}
```

2.6 A Java API for MDSplus

In order to make progress in constructing EScope, we will need to have a Application Programming Interface (API) for accessing MDSplus. On the accompanying CD, we have provided three helper classes and one interface which will be all that you need in order to download and visualize typical data entries. Their full listings are provided in Appendix C. (These classes are a simplified version of the Java tools provided with the full MDSplus package. We admit that their structure is based on the historical development of MDSplus rather than being excellent examples of object-oriented design!)

The helper classes comprise the following:

- MDSDescriptor is a class which stores and describes a single MDSplus dataset. It contains a predefined set of constants which are used to flag the type of data. It also contains the data array itself and a number of methods for storing and accessing it. This class is a simplified version of the information which is stored at every data-containing node in the MDSplus tree. The beginning of this class is shown in Listing 2.5.
- MDSMessage is the class which negotiates remote access to MDSplus. Using the *mdsip* protocol described in Section 2.7.1, string "messages", which consist of a packet of header information and an expression to be evaluated, are sent to the server. The MDSplus server evaluates the expression and the result is received by MDSMessage, decoded into a MDSDescriptor and eventually returned to a client.
- MDSNetworkSource wraps up the negotiation with the MDSplus server into high-level methods like connect, open, evaluate and so on. All a client

program needs to know about are the method signatures for the some of these methods, so it makes sense to have this class implement an interface, `MDSDataSource`, which, in principle, is provided to the client software.

The way in which these classes work together is shown schematically in Fig. 2.6. Because data is eventually returned to the client program wrapped up in a `MDSDescriptor`, the client needs to know about this class as well as the `MDSDataSource` interface. The meaning of the symbols in this figure will be described later in this book.

The following sections will discuss further details of the data organization and the remote-data-access protocol of `MDSplus`. Their aim is to more fully motivate and describe the helper classes. If you have made a start reading the `MDSplus` web-site then you might find this discussion useful. For many readers, they can be skimmed over on a first pass (perhaps dwelling briefly on the discussion of `MDSNetworkSource` in Section 2.7.3).

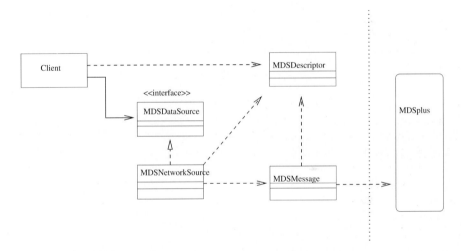

Fig. 2.2. Schematic representation of the helper classes `MDSDescriptor`, `MDSMessage`, `MDSNetworkSource` and the interface `MDSDataSource` showing the way in which they interact with a client class and the `MDSplus` database.

Listing 2.5. Part of the `MDSDescriptor` class used to hold data returned from MDSplus.

```
/** Used to store the response from the MDSPlus server */
public class MDSDescriptor
{    // Predefined constants flag the type of data:
    public static final byte MAX_DIM       = 8;
    // string or character data
    public static final byte DTYPE_CSTRING = 14;
```

```
        public static final byte DTYPE_CHAR    = 6;
// integer data
        public static final byte DTYPE_BYTE    = 2;
        public static final byte DTYPE_SHORT   = 7;
                //(short is converted to int)
        public static final byte DTYPE_INT     = 8;
// floating point data
        public static final byte DTYPE_FLOAT   = 10;
                //(float is converted to double)
        public static final byte DTYPE_DOUBLE  = 11;
// unsigned word
        public static final byte DTYPE_WORDU   = 3;
                //("usigned word" is converted to int)
// event
        public static final byte DTYPE_EVENT   = 99;

// The following variables store the data:
        private byte     descriptorType;

        private byte     byteData[];
        private int      intData[];

        private double   doubleData[];

        private String   charData;
        private String   cstringData;
        private String   eventData;
        ....
}
```

2.7 The Data Organization of MDSplus

In this section we will briefly discuss the structure of the MDSplus database as well as the APIs for accessing and manipulating data. Readers may also wish to consult Appendix A for more information about installing and running MDSplus.

MDSplus assumes that the top level description of data is that of an "experiment". Experiments can be defined in the database to be comprised of a number of files corresponding to data recorded from physical diagnostics and control parameters. This files are arranged in a tree hierarchy which is something which suits fusion experiments which have a large number of diagnostics concerned with notions such as "radio-frequency power", "neutral-beam injection", "coil currents", "spectroscopy" and so on. A physical diagnostic within one of these categories might measure a number of signals so it is easy to imagine that a tree structure for the entire database would be 3 or 4 levels deep.

2.7 The Data Organization of MDSplus

Each data record for an MDSplus experiment is given the name of a "shot" or "pulse". This is the popular jargon for a single plasma discharge: the ionized gas is pumped into the chamber, the magnetic coils are energized and the gas is heated to temperatures approaching those of the sun. Many diagnostics collect megabytes of data over periods of time ranging from some tens of milliseconds to several seconds. Many shots are taken during one experimental campaign.

Every dataset in an MDSplus experiment is identified by its path inside the tree structure. For example

.OPERATIONS:I_FAULT

specifies the i_fault (fault current) dataset which sits under the operations sub-tree (of the h1data experiment) Note that MDSplus names are case-insensitive and they begin with either a "." or a ":". It is also possible to associate a unique name, called a *tag*, with a dataset in order to simplify identification. Tags are a convenience mechanism and are often defined for the most frequently-accessed data, such as significant experimental results.

In MDSplus trees the leaves typically store data and internal nodes are typically used to specify the hierarchy without storing data. A dataset may specify a variety of information, ranging from string expressions to scalar values to scalar arrays to arrays composed of a sequence of pairs (with each pair describing a physical measurement and the sampling time). A data descriptor is associated with every data item in an MDSplus experiment and contains the definition of the the data type, its dimension and other information.

As mentioned above, EScope uses a simplified data interface to MDSplus and handles a subset of the MDSplus data types. Because, in EScope, we will be interested only in those data items which describe the evolution of some quantity over time, we will only need to handle arrays of data values and strings to describe data and error messages.

It often happens that a scientific user is not interested in a simple data item but, rather, in a combination of different items. For example, it might be the case that a user requires the values of one dataset to be scaled or offset by another: a data reference in a shot file my_data might need to be processed as

my_data_gain * (2.3 * my_data) − my_data_offset

For this reason, data access in MDSplus is, in general, carried out by providing generic *expression evaluations*. A specialized language, TDI ("Tree Data Interface"), is provided for this purpose. Expressions may also specify that a specific user-supplied analysis program gets executed at the time the expression is evaluated.

A detailed description of the TDI expression syntax is outside the scope of this book but it is important to realize that every data access in MDSplus corresponds to the evaluation of some kind of expression. The methods provided in the helper classes therefore define a string input argument representing the expression to be evaluated (possibly a simple reference to a data item) and return a descriptor object which describes the result of the evaluation.

2.7.1 The *mdsip* Protocol for Remote Data Access

The following discussion of the `mdsip` *protocol for remote access to* `MDSplus` *can be skipped or skimmed on a first reading. It complements the discussion of* `mdsip` *in Appendix A.*

Many modern database systems provide for remote data access, i.e. accessing the database from a machine which is different from that hosting the database itself over a network connection. Remote data access requires the following components:

- A database server, i.e. an application running on the machine hosting the database and accessing data on behalf of the client currently requesting the data.
- A client API, usually implemented as a library. The definitions of the methods defined in the API are usually close to the methods defined for local database access. The implementation of these methods handle network communication with the database server. Typically, a client application first connects to a database server. The established network connection remains active until the client disconnects from the database server or exits.
- A network protocol for data exchange. The client and the server need to communicate properly. Most database applications use the TCP/IP protocol for communication. TCP/IP allows the establishment of a connection, and reliable communication over it. Using a network protocol which guarantees reliable communication has the advantage that the code that handles the client-server communication does not need to handle the many problems which arise in network communication such as loss of data or the corruption of data packets. It is, nevertheless, necessary to build an application-specific protocol over the network layer.

The remote-data-access layer of `MDSplus` defines its own protocol over TCP/IP, called `mdsip`. In a data-access transaction, the client sends a string specifying the expression to be evaluated and the `mdsip` server returns the result of the evaluation. As different kinds of data can be returned by expression evaluations, `mdsip` specifies the transmission of a descriptive header, followed by the data itself. The client can handle the reception of a different number of bytes because this information is available after the fixed-length header has been received. (In TCP/IP it is necessary to know the number of bytes being received in advance since this information is not provided by the network layer.)

The `MDSMessage` class, defines all the information required for sending an expression as a string to the server and retrieving the result of the evaluation. The "packet" exchanged via TCP/IP with `MDSplus` contains the following information:

- `msglen`: the total length of the message in bytes.

- `status`: the status of the expression evaluation. This field is valid only when the message carries the result of the expression evaluation, i.e. when it is transmitted by the `MDSplus` server back to the client.
- `length`: the length in bytes of the *data field* included in the message. The total length of the message is, therefore, the sum of the data length and of the length of the associated header (which is 48 bytes).
- `nargs`: the number of arguments. The expressions being evaluated are described as strings, but it is possible to define additional arguments, which are sent by the client to the server and then used in the expression evaluation. This feature is useful, for example, when the `mdsip` protocol is used for application servers in which powerful computers are used to run simulation codes to model experimental results. In this case the expression being evaluated can invoke a separate simulation program. It is interesting to observe that the syntax can generalize to cases where the arguments might contain a huge string of data values: The arguments can be identified in the expression by the symbols $1, $2, ..$n, and the message will be followed by other messages containing a binary encoding of the arguments. This is beyond the scope of this book.
- `descr_idx`: the "descriptor index". As noted above, when arguments are defined in the expression to be evaluated, more `mdsip` messages may need to be sent by the client. The server knows the number of arguments it is going to receive after receiving the fist message. The `descr_idx` labels the subsequent arguments. This is beyond the scope of this book.
- `dtype`: the type of the data associated with this message. In `EScope` only string messages will be sent to the server so `dtype` will always be `DTYPE_CSTRING`. On return from the server, `dtype` can take the range of values shown in the `MDSDescriptor` class.
- `client_type`: the "type" of the client computer. This is to register the "endianness" and IEEE floating point encoding of the client computer with the `mdsip` server. Because the byte order of data storage varies between computer manufacturers, it may be the case that bytes transmitted in "big endian" format over a network need to be "swapped" for the client machine. When the `mdsip` server receives the `client_type` flag, it decides whether to swap bytes for transmitting the result of the expression evaluation. At the time of writing, `mdsip` violates the convention that data transmission over TCP/IP should be big endian. Instead it determines the endian type of the client machine and modifies the data accordingly. This may result in an efficiency gain under some circumstances. (The `client_type` flag also determines whether any translation of floating point format needs to be done before transmission of data by the server. We shall hide its complexities inside the `MDSMessage` class and forget about it. Such is the situation with some real-world case studies!)
- `msgid`: an identifier for the message which is copied by the `MDSplus` server onto its reply. This identifier can be used to label messages in, for example, situations where a threaded client might be making several requests in

different threads, possibly with several `mdsip` servers. In this case the `msgid` field can be used to properly associate answer with requests.
- `ndims`: the number of dimensions in an array of binary data. We will not make use of this field in our `EScope` case study as we will only be concerned with one-dimensional arrays.
- `dimensions`: a fixed size array of integers specifying each dimension (up to eight dimensions). This field and the the previous one are used to handle multidimensional data, serialized in row-first order in the associated data buffer. Once again, this will not be relevant to our `EScope`.
- `body`: the data buffer containing serialized data. The reconstruction of the transmitted data from the content of this buffer is unambiguous when combined with the contents of the `mdsip` header (particularly `dtype`, `ndims` and `dimensions`).

2.7.2 Operation of `MDSMessage`

The above variables define the `mdsip` protocol, being the set of rules the client and server must adhere to to communicate unambiguously. The constructor of `MDSMessage`, shown in Appendix C.4, accepts a string argument and constructs a `mdsip` header corresponding to a request to evaluate the expression corresponding to this argument. The method `send()` had an instance of `DataOutputStream` as its argument. Using a Java `DataOutputStream` for sending messages, and a `DataInputStream` for message reception, simplifies the serialization process because `DataOutputStream` provides the correct serialization in its `write...` methods. The `mdsip` protocol defines twos-complement and IEEE 754 formats to serialize integer and float numbers, respectively. These formats are supported by the `DataInputStream` and `DataOutputStream` classes in their `write..` and `read..` methods.

By way of illustration, the `MDSMessage` method `receive(DataInputStream s)` reads the incoming `mdsip` message from the `DataInputStream` instance "s" and performs the following steps:

1. Fill in the `MDSMessage` fields from the fixed length header. This operation is done in two steps: first a fixed-length byte array corresponding to the bytes of the message header is read from the input stream. Then the desired fields are retrieved, using the support methods `byteArrayToInt`, `byteArrayToShort` and `byteArrayToShort`.
2. Read the rest of the message. When the header has been decoded, the number of incoming bytes is known and it is possible to decode them correctly, based on the knowledge of the data type and of the dimensions.

Once the incoming message has been decoded, the received data, converted into Java data types, is returned by method `receive`. The result is encapsulated into an instance of the `MDSDescriptor` class. This class describes a generic scalar or array data item and resembles the descriptor structure defined in `MDSplus`.

2.7.3 Operation of MDSNetworkSource

As we have just seen, MDSMessage is responsible for the proper management of the mdsip network protocol once a proper network connection has been established. Managing the connection to MDSplus is done by the MDSNetworkSource class whose public interface has the following methods:

- connect(String serverAddrCPort): represents the first action carried out to establish a connection. The string argument specifies the IP address and the port number of the mdsip server (separated by a colon in the string). The method establishes a socket and input and output streams as we did in our earlier SimplePlotClient example.
- open(String experiment, int shot): opens the specified experiment and shot database. This is done by sending a special expression JAVAOPEN(experiment,shot) to MDSplus.
- evaluate(String expression) calls the MDSMessage send and receive methods to evaluate the string expression in its argument. The result is returned as a MDSDescriptor object.
- close() closes a currently opened shot database.
- disconnect() terminates a currently established network connection to MDSplus.

At any time a connection to a mdsip server is in one of the following states:

- *initial*: no connection has been established.
- *connected*: a connection has been established, but no shot database has been opened.
- *open*: a shot database is currently open.
- *closed*: a previously opened shot database has been closed and no other database is now open.
- *disconnected*: a network connection to a mdsip server has been closed and no other connection is currently active.

The behavior of the methods of MDSNetworkSource depends on the current state of the connection. For example, the method evaluate can be successfully executed if and only if the connection is open. The implementation of this method checks the status of flags isConnected and isOpen to determine whether or not the expression can be sent. Exceptions are thrown if it cannot be sent. We will return to the structure of this class as our case study evolves.

2.8 PreEScope0: A Program to Connect to MDSPlus

We would like to propose that this section be accomplished as an exercise by the reader. It is not "compulsory" and you could turn to Appendix D to find the solution if you wished. But it is only a small step beyond what you have learnt from the earlier examples in this chapter and you can use the supplied helper classes to connect to the `MDSplus` database or to the simulator program described in Appendix A.1.

The specifications of this program are as follows:

1. It will run in the command line and take four arguments: the server address and port (in the form "address:portNo"), the name of the experiment, the shot number and the name of a dataset to be downloaded.
2. After checking that the arguments have been supplied correctly, the program will open a connection with the `MDSplus` database and download the data corresponding to the particular dataset. One example dataset is the ".operations:i_fault" leaf node.
3. The X and Y data arrays of this dataset are then written to two binary files `data_yVals` and `data_xVals`.
4. The `Plotter.plot` method will be called to plot the data.

Does this sound familiar? In fact, this program will set up the files that you need to run `SimplePlot`. And you can run `SimplePlot` to test it out.

You can also try sending other commands to the `MDSplus` server and printing out its response. For example

- Sending "10+2" should return the number "12".
- Sending "units_of(.operations:i_fault)" should return a string "Amps".
- Other datasets such as .operations:diamag, ... can be downloaded, plotted up and written to file.

2.9 Programming Exercises

The following exercises will help readers become familiar with the parts of the Java API described in this chapter.

1. As well as the binary input classes, Java has a completely analogous set of classes to *output* data to binary files. You can look these up in Sun's Java API documentation and then try modifying the `SimplePlot` program to write out the data arrays into different files. You can then read the data back in from these new files to make sure that they have been written correctly. You could make this slightly more complicated by writing the data into files in different directories on your computer.
2. You could now try to modify `SimplePlot` to read text data. Just type some X and Y data into two text files using an editor. Calling your modified `SimplePlot` will now construct an unlabelled line graph of the data you typed! Modify the data to draw some interesting waveform shapes.
3. You could now write a program to solicit a number of (X,Y) data pairs from a user and then plot them up.

4. If you have already converted `SimplePlot` to read text data, you will be forced to handle `NumberFormatExceptions` when you convert this data to Java primitive types. You might wish to practice catching these exceptions and testing your program with good and bad data.
5. Download the `SimplePlotServer` examples from the CD to your computer. You run them by starting the server from a different terminal window to the client or by running it in background from the same terminal window. Play with the examples and verify that the commands can be sent in a different order from the client to the server. You might then like to set up your own client/server system using a simple language to talk between them:

   ```
   CLIENT: Hello
   SERVER: Hello. I am a server. How can I help?
   CLIENT: Time
   SERVER: 1126870937252
   ```

6. The `SimplePlotServer` program exits if a client sends a `CLOSE` command or a null line. You might think that it would handle multiple clients if it were modified to have an infinite server loop. This turns out not to be the case and, as described in the next section, you need to turn to Java threads to build a true, multiple-client server.

   Try out making the loop infinite by putting a line "`done =false;`" at the end of the `SimplePlotServer` while-loop. If you start one client then you will see the graph. If you start another client then you will not see another graph and the client will hang. Remember to stop the server as well, by crashing it, when you have finished with this exercise. (On Linux, you can find the process number of the server by typing `ps` and looking for `java`. You can "kill" this process by typing `kill -9` followed by the process number.)

2.10 Further Reading

Sun's web documentation on the Java API [12] and Java Tutorial [13] are book-marked by every Java programmer.

There are many Java programming texts which are suitable for background reading about Java IO. We particularly acknowledge, and recommend, those by Hunt ([14]), Horstmann ([15]) and Horstmann and Cornell ([16]).

3
Graphical User Interfaces Using Swing

Java has very extensive graphics capabilities. You use Java for

- building graphical user interfaces, or GUIs, (using Swing)
- 2D graphics (using Java2D)
- 3D graphics (using Java3D [17] or JOGL [18])

The basic Java graphics package is called the Abstract Window Toolkit (AWT). A great deal of work has been put into writing graphics packages which improve the capabilities of this toolkit. Some Java text-books make extensive comparisons between the more modern packages and AWT. This can become annoying if all you want to do is just use the modern packages. But, unfortunately, it is not possible to just forget about AWT and use the new packages because, for example, many Swing classes are descended from AWT classes such as `java.awt.Component`. Additionally, the event-handling mechanism for Swing components, discussed in Section 3.4 below, actually comes from AWT.

Modern graphics and GUI's are very successful applications of object-oriented programming and a study of them *might help* you feel more comfortable with the whole OO philosophy. Graphics and GUI programs have patterns (both formal design patterns and informal patterns) which are often repeated. If you learn these patterns then you can be relaxed about many of the details.

3.1 Simple GUI Programming

3.1.1 A Blank Frame

Study the pattern of the following code which creates a blank window:

Listing 3.1. A simple blank GUI frame.

```java
import javax.swing.*;
/** Creates a blank window.*/
public class BlankFrame extends JFrame
{
    public BlankFrame() {}
    public void initialise()
    {
        setSize(600,400);
        setLocation(200,200);
        setDefaultCloseOperation(JFrame.EXIT_ON_CLOSE);
        setTitle( "EScope: Blank Frame" );
        setVisible(true);
    }
    public static void main(String[] args)
    {
        BlankFrame ourEScope = new BlankFrame();
        ourEScope.initialise();
    }
}
```

There are some features of this example which deserve comment. `BlankFrame` extends the Java Swing `JFrame` class. It contains calls to `JFrame` methods to set the size of the frame, to position it on the screen, to set its title and, finally, to display it on the screen (using `setVisible(true)`).

The `main` method constructs a `BlankFrame` object, calls its `initialise` method and then exits. The thread controlling the Swing window remains active even after the main method has finished. The window displayed will eventually become a GUI which is able to accept and process user events such as button clicks. The Swing thread will terminate if someone clicks on the "close-window box" in the top, right-hand corner of the frame. This closing behavior was specified by calling the `setDefaultCloseOperation` method of `Jframe`.

`JFrame` has methods (inherited from `java.awt.Component` and `java.awt.Window`) to resize and position the frame. For example:

```
setSize(WIDTH, HEIGHT);
setLocation(x,y);
setBounds(x, y, WIDTH, HEIGHT);
```

The method `setLocation(x,y)` will move the frame to x pixels to the right and y pixels down the screen from the top left corner. It is possible to obtain the dimensions of your screen using a class called `java.awt.Toolkit`. Using this class you can ensure that your frame is in the middle of any given screen. If you do not want specify your frame size, there is another method, called `pack()`, which can be called to make sure that all of your components are visible.

Frames are examples of *containers* and we will study ways of inserting graphics objects into them in the next example.

3.1.2 Laying-out Components in a `JPanel`

Many Swing GUI objects are able to display other GUI objects. Most importantly, a `JFrame` will end up being a container for buttons, menus, labels, graphics and so on. These are drawn onto the "content pane" of the `JFrame` and you need to obtain this object using a method `getContentPane()`. You then need to "add" the GUI components to this content pane.

Actually, this informal pattern has one more layer of complication: by convention it is preferred to draw graphical objects onto a "panel" container object and then to add this panel object to the base frame. This is illustrated in the next listing. Two views of the GUI are shown in Figs. 3.1 and 3.2.

Listing 3.2. A frame containing a number of Swing components.

```java
import javax.swing.*;
/** Places a number of Swing GUI components onto a frame
    using flow layout.*/
public class ComponentFrame extends JFrame
{
    private JPanel componentPanel;
    public ComponentFrame()
    {
        // Components for the GUI
        JMenuBar menuBar = new JMenuBar();
        JMenu menu = new JMenu("Drop Down Menu");
        JMenuItem item = new JMenuItem("Menu Item");
        JButton aButton = new JButton("a button");
        JButton aNButton = new JButton("another button");
        JButton yANButton = new JButton(
                              "yet another button");
        JLabel aLabel = new JLabel("label");
        JTextArea text = new JTextArea("Type here!");
        JRadioButton radio = new JRadioButton();
        JCheckBox check = new JCheckBox();
        JSlider slider = new JSlider();
        JComboBox combo = new JComboBox();

        menuBar.add(menu);
        menu.add(item);

        componentPanel = new JPanel();
        componentPanel.add(menuBar);
        componentPanel.add(aButton);
        componentPanel.add(aLabel);
        componentPanel.add(aNButton);
```

```
        componentPanel.add(text);
        componentPanel.add(radio);
        componentPanel.add(yANButton);
        componentPanel.add(check);
        componentPanel.add(slider);
        componentPanel.add(combo);
    }
    public void initialise()
    {
        getContentPane().add(componentPanel);

        setSize(600,400);
        setLocation(200,200);
        setDefaultCloseOperation(JFrame.EXIT_ON_CLOSE);
        setTitle( "EScope: Component Frame" );
        setVisible(true);
    }
    public static void main(String[] args)
    {
        ComponentFrame ourEScope = new ComponentFrame();
        ourEScope.initialise();
    }
}
```

Fig. 3.1. GUI for the `ComponentFrame` example described in the text. A radio-button is on the right hand side of the top row. A check-button is in the middle of the second row next to the slider bar. A combo-box is on the right of the second row.

Fig. 3.2. The GUI for the `ComponentFrame` example described in the text. The radio-button and check-button have been checked and the slider has been moved.

3.1 Simple GUI Programming

In this example, the constructor creates a `JPanel` object and then adds a number of GUI components to it. The `initialise` method has a line to call `getContentPane` and to add the `JPanel` object to the content pane.

This example does nothing useful because we have not hooked up any of our own event-handling code to handle mouse clicks. But some of GUI objects still have a reaction to mouse events. You can have some "serious fun" by clicking on the menu, the buttons, the radio button, the check box, the combo box and the slider bar. Notice how the "focus" of the GUI changes as you activate the various components. GUI objects with focus are sometimes visually highlighted and can respond to keyboard events as well as mouse clicks. For example, if you give focus to the slider bar you can then move it using the arrow keys on the keyboard.

The text-area component of our example is even more fun! Start typing text into it and observe how it expands in width. As it expands, it, and all of the GUI objects after it, *flow* around the borders of the frame. This is because the default "layout" for `JPanel` objects is "flow layout".

Using a careful combination of the various layouts available in Java, it is possible to position GUI components precisely and attractively. The *grid layout* can be used to locate components in a regular matrix of locations. You can construct your own `GridComponentFrame` by modifying `ComponentFrame` so that it imports the `java.awt.GridLayout` class. Then you can set the layout after constructing the panel:

```
componentPanel = new JPanel();
componentPanel.setLayout(new GridLayout(2,5));
```

Border layout divides a panel into five fields known as "North", "East", "South", "West" and "Center". The actual shape and size of these fields depends on which has been populated with GUI objects. It is worth experimenting with the example in Appendix B.1, on page 243, to see how this works. The default field is "Center" and if you keep adding GUI components then they end up sitting on top of one another in the central position. Just as flow layout is the default for a panel, border layout is the default for a frame.

3.1.3 A Bizarre Component Frame

Finally, we would like to present a bizarre example in which some GUI objects are added directly on top of other objects. Notice the buttons inside a button and the slider inside a menu! Constructing strange interfaces such as this one is a good way to get the feel for the way in which Swing objects hook together. There is much to be learnt by experimenting and having fun!

Listing 3.3. A bizarre example which shows that Swing components can be added on top of one another.

```
import java.awt.GridLayout;
import javax.swing.*;
/** A grid layout example with a bizarre arrangement
```

```java
                of GUI components. */
public class BizarreComponentFrame extends JFrame
{
        private JPanel componentPanel;

        public BizarreComponentFrame()
        {
                //Components for the GUI
                JMenuBar menuBar=new JMenuBar();
                JMenu menu = new JMenu("Drop Down Menu");
                JMenuItem item = new JMenuItem("Menu Item");
                JButton aButton=new JButton("a button");
                JButton aNButton = new JButton("another button");
                JButton yANButton = new JButton
                                        ("yet another button");
                JLabel aLabel=new JLabel("label");
                JTextArea text = new JTextArea("Type here!");
                JRadioButton radio = new JRadioButton();
                JCheckBox check = new JCheckBox();
                JSlider slider = new JSlider();
                JComboBox combo = new JComboBox();

                menuBar.add(menu);
                menu.add(item);
                menu.add(slider);

                aButton.setLayout(new GridLayout(5,2));
                aButton.add(aNButton);
                aButton.add(yANButton);
                radio.add(text);

                componentPanel = new JPanel();
                componentPanel.add(menuBar);
                componentPanel.add(aButton);
                componentPanel.add(aLabel);
                componentPanel.add(text);
                componentPanel.add(radio);
                componentPanel.add(check);
                componentPanel.add(combo);
        }
        public void initialise()
        {
                getContentPane().add(componentPanel);

                setSize(600,400);
                setLocation(200,200);
                setDefaultCloseOperation(JFrame.EXIT_ON_CLOSE);
                setTitle( "EScope: Bizarre Component Frame" );
                setVisible(true);
```

```
    }
    public static void main(String[] args)
    {
        BizarreComponentFrame ourEScope =
                            new BizarreComponentFrame();
        ourEScope.initialise();
    }
}
```

3.2 A Note on Programming Style

In this book, we make no claim that our example programs display the best-ever Java coding style, but we hope that they are reasonably clear and consistent. Our style convention is based on a mixture of those adopted by Lewis and Loftus [19] and Horstmann [15].

We like corresponding curly braces, which define block structures in Java, to have the same indentation. This is not the preferred Sun style [20] but it is sometimes used in introductory Java texts. For publication purposes, we sometimes break this matching-braces rule occasionally when a method, or a statement, is trivial enough to be written on one line. (We have also broken the rule by mistake in some of the listings on the CD because modern editors, especially the free ones we use, like to take control of indentation and sometimes the result is not what we intended!)

We have tried to be reasonably strict about keeping the class instance variables **private** in order to enforce information-hiding (although some **protected** variables will sneak in later in Chapter 10).

Some of our other conventions are not so strict and we probably should have more comments. Our lack of systematic commenting has been done, in part, to enhance the readability of printed listings in this book. For example, we usually use short, single-line comments rather than long **javadoc** comments, to enable printed code listings to be easily read. (There does seem to be a conflict between placing systematic **javadoc** comments, to enable hypertext browsing of Java software, and having a listing which is pleasant to read in a printed format or at a terminal!) In general, the issue of good commenting in software is addressed in many Java text books. Readers are welcome to adopt their own style conventions by consulting text-books, by reading code written by colleagues, by adopting conventions forced by Integrated Development Environments or by exercising a programmer's right to be an individual. But it is quite important to reflect on your coding quality and to be as clear as you can. Remember that the poor person who has to read your code in many years time might well be yourself!

Our examples in this chapter are slightly more complicated than they really need to be. Take, for example, **BizarreComponentFrame**, in Listing 3.3. It would have been possible to have made this "simpler" by writing the entire code inside one `main` method. This would have been considered poor programming style – `main` methods are meant to be used to just "drive" an object-oriented system and they should not be over-burdened with code. The code which is specific to the classes themselves is meant to be located within the class methods (including their constructors). It would also have been possible to simplify this example by eliminating the `initialise` method and placing all of the frame-specific code inside the `BizarreComponentFrame` constructor. We have, instead, adopted an "informal pattern" which splits up the coding which would normally be associated with a constructor method into two parts: method calls which specify the *appearance of the frame and its components* will generally appear in a constructor. Methods which *involve hooking the class up to other classes, and which finally position and display the frame*, will appear in an `initialise` method. The idea is that some amount of initialisation is best placed inside a constructor, but other initialisation tasks are best carried out using a constructed object. It is hard to justify this convention at this stage in the book, but, readers might agree with it after following the development of the `EScope` case study from Chapter 8 onwards.

As a final point, we note that some authors advocate having a `main` method inside every class. In this way, classes can be tested and debugged separately before hooking them up to the main method of your entire program. This practice emphasizes the reusability of Java classes. But it does have a trap if you want to create a new class by copying an existing one: you need to be careful that the main method for the new class will instantiate an object of the new class type. Otherwise, if they are in the same directory on your computer, your new class might compile and execute but just instantiate an object of the old class type that you copied! (Believe us, these problems can be very frustrating to debug!)

3.3 A Look Inside `Plotter`

Up until now, we have deliberately shielded our readers from the inner workings of the `Plotter` class which we used to plot up waveform data in Chapter 2. There is much in this class which has to do with `Java2D` graphics which is the topic of the next chapter. But, in the end, graphics are drawn onto a panel which becomes part of a Java GUI. So it is appropriate to start peeping inside this class now and you are welcome to have a peek at Appendix B.2 on page 244. Its basic structure is similar to the simple examples we have already met in this chapter. The constructors do nothing except instantiate objects. The basic frame is initialized in an `initialise` method and the data is passed to a panel by calling a `setGraphData` method. Eventually `EScope` will follow this informal pattern.

Recall that our examples in Chapter 2 took arrays of X and Y data and plotted them up in a separate window. To do this a call was made to the static method `plot`:

```
Plotter.plot(xVals,yVals,"test");
```

In order to draw graphics on a panel, the `plot` method needs to construct a panel which overrides a method of the `JPanel` class called `paintComponent`. This is a very unusual method. Consider the following points when reading the listing in Appendix B.2 (page 244):

- You never call `paintComponent`! It is automatically called by the operating system "when it is needed". You never call it yourself but you can force the operating system to "repaint" your graphics by calling the `repaint()` method of the `JComponent` object which contains your graphics `JPanel`.
- `paintComponent` takes a single argument of type `Graphics` which remembers the current state of the graphics on your panel. The state includes information such as the present text font and the present drawing colour. The `Graphics` object also has methods which are used to draw lines, images, and text. Remember that you do not pass this `Graphics` object to the `paintComponent` method at all; the operating system does.
- Inside the `paintComponent` method, you need to paint the panel itself and fill it in with a background color by calling the `paintComponent` method of its parent object:

  ```
  super.paintComponent(g);
  ```

- In order to use the `Java2D` API the `Graphics` object needs to be cast to a `Graphics2D` object.

More details of this `paintComponent` method will be discussed in Chapter 4 below.

3.4 Action Listeners in Swing

In order to have your graphical interface do anything useful, there has to be some way of recognizing and acting on the *events* that occur when a user performs a mouse or keyboard operation. As mentioned in Section 3.1.2, some Java GUI components have automatic event recognition built into them. For example, if you clicked on the `JMenu` component of the `BizarreComponentFrame` example you would have noticed that it became highlighted and dropped down to display its menu items. For another example, if you clicked on a `JButton` with the left mouse button, and held your finger down, then the `JButton` became highlighted. Releasing the mouse button would have taken the highlighting away to reveal a small box around the `JButton`'s title; this box indicates that the `JButton` has *focus*. `JButton`s with focus can be activated again by pressing the space bar on your keyboard.

To build a useful application, you need to augment these in-built component behaviors by writing classes to define *listener* objects which can perform custom *actions*. There are several ways which this can be done. We shall illustrate some of them below and elsewhere in this book. Because Java is an object-oriented language, it turns out that events and listeners are objects. The actions to be performed are defined as methods inside the listener classes. These methods get called by the operating system when an event is registered and they accept the event object as an argument. You define your listener classes by implementing interfaces and extending classes from the `java.awt.event` package. Listeners in Swing are an example of the "observer" design pattern discussed in Chapter 14.

In the next example, a listener class is defined to implement a "Quit" menu item. A listener object is constructed and added to that menu item. The listener object then reacts when the menu item is selected. The listener class implements the `java.event.ActionListener` interface which contains one method: `actionPerformed(ActionEvent e)`. In this example the listener class is an *inner class* of the main frame. Inner classes are completely contained in another class. They can be made *private* so that they are invisible outside of that class just like private data fields. They can access the fields and methods of the enclosing class so they are useful if your `actionPerformed` method needs to change the appearance of the GUI object. The main drawback of inner classes is that they can clutter up your code, especially if the `actionPerformed` method is lengthy and especially if you need to have several inner classes defined inside one enclosing frame. Our recommendation is to be careful with too much use of inner classes. They have a deserved place in user-interface programming but you might consider delegating your action processing elsewhere if it is becomes very complicated. Later in this book you will see how the `EScope` case study delegates processing away from the GUI classes.

Listing 3.4. A class which contains an `ActionListener` inner class.

```java
import javax.swing.*;
import java.awt.event.*;
/** A simple listener example which implements
    an "Exit" menu item.*/
public class SimpleExitListener extends JFrame
{
    private JPanel componentPanel;

    public SimpleExitListener()
    {
        //Components for the GUI
        JMenuBar menuBar=new JMenuBar();
        JMenu menu = new JMenu("File");
        JMenuItem item = new JMenuItem("Exit");

        menuBar.add(menu);
        menu.add(item);
        setJMenuBar(menuBar);
        item.addActionListener(new ExitAction());
```

```java
        componentPanel = new JPanel();
    }
    public void initialise()
    {
        getContentPane().add(componentPanel);

        setSize(600,400);
        setLocation(200,200);
        setDefaultCloseOperation(JFrame.EXIT_ON_CLOSE);
        setTitle( "EScope: Simple Exit Listener" );
        setVisible(true);
    }
    public static void main(String[] args)
    {
        SimpleExitListener ourExitListener =
                                    new SimpleExitListener();
        ourExitListener.initialise();
    }

    private class ExitAction implements ActionListener
    {
        public ExitAction() {}

        public void actionPerformed(ActionEvent event)
        {
            System.exit(0);
        }
    }
}
```

Our listener examples can be made significantly more fun and interesting by changing some aspect of the appearance of the GUI. This is provided in the `SimpleExpandAction.java` program on the accompanying CD. In this program, "expand" and "contract" menu items are added to change the size of the base frame. The following listing shows the listener classes for this example.

Listing 3.5. These listener classes change the appearance of the GUI.

```java
    private class ExpandAction implements ActionListener
    {
        public ExpandAction() {}
        public void actionPerformed(ActionEvent event)
        {
            setSize(600,600);
        }
    }
    private class ContractAction implements ActionListener
    {
```

```
    public ContractAction() {}
    public void actionPerformed(ActionEvent event)
    {
        setSize(300,300);
    }
}
```

3.5 Swing Miscellany

There is a lot of detail in the Swing API. Whole books have been written about it. The Sun tutorial pages provide extensive examples to help you get started. We would like to mention some miscellaneous aspects of some of these components here. Our introductory `ComponentFrame` example, shown in Figs. 3.1 and 3.1, displayed some of these. We will deal with some more complicated aspects of Swing as our `EScope` case study develops.

Labels

Labels display text. Some Swing components, such as `JButton`s, have their own labels. Others can be labelled using the `JLabel` component. `JLabel`s can also be used for communicating information to the user. The `Jlabel` constructor has fields for specifying the initial text as well as the text alignment (`Jlabel.LEFT`, `Jlabel.CENTER` or `Jlabel.RIGHT`). It is possible to specify HTML tags in the label so that it can be beautifully formatted.

Check Boxes and Radio Buttons

The check box in our `ComponentFrame` example appeared as a square with or without a "tick" depending on whether it had been checked or not. Check boxes are usually used to select a number of "yes" or "no" options from a list. Each list item would comprise a `JLabel` describing the option followed by a check box. It is a convention that several check-box alternatives can be specified from the list at the same time.

Radio buttons appear as a blank circle with or without a solid disk inside depending on whether or not they have been checked. Radio buttons are usually used to select one alternative from a list in an *exclusive* way. To achieve this, `JRadioButton` objects need to be added to a `ButtonGroup` object. The `ButtonGroup` enforces the "only-one-radiobutton-on" protocol by keeping track of the states of all of the buttons in the group. Our `ComponentFrame` example only had one radio button so there was no need to group it.

ComboBox

Combo Boxes supply a drop-down menu with, optionally, an editable window in which the current selection is displayed. You can change the current selection by editing the window or by selecting one of the fixed number of options from the list. The text that you enter can be different from the other options on the list but it must be a legal option for the program. Our example did not have an editable window.

Sliders

Sliders are attractive ways of allowing users to select from a range of values. They can sit horizontally or vertically. You retrieve their values by attaching a listener object which implements the `ChangeListener` interface. This has one method, `stateChanged`, which is used as follows:

```
public void stateChanged ( ChangeEvent event )
{
    JSlider slider = ( JSlider ) event . getSource ( ) ;
    int value = slider . getValue ( ) ;
    ...
}
```

There are wonderful things that you can do to the appearance of sliders including adding tick marks, snapping the selection to a particular tick, filling in part of the slider area and adding labels and icons.

Menus

In order to build a conventional menu, you create a menu bar and add it to a panel. You then create a menu object and add it to the menu bar. Then create menu items and add them to the menu. This was demonstrated in our `ComponentFrame` example.

You need to install an action listener for each menu item. It is possible to add icons to menu items and to include radio buttons and check boxes in menu items. Menus can be changed dynamically at run-time. In particular, it is possible to *disable* menu items. Disabled items cannot be selected and will appear to be "greyed-out".

The Swing API also includes functionality for specifying keyboard commands as short-cuts to menu item selection (Alt-key combinations and Crtl-key combinations).

It is also possible to have pop-up menus. These are not tied to a menu bar but pop up when you click a certain mouse button (typically the right mouse button). You need to install a mouse listener which we will discuss in Chapter 5.

Actions

Instead of implementing the `ActionListener` interface it is possible to extend the abstract `AbstractAction` class.

The abstract `AbstractAction` class implements several interfaces. Of these, only `ActionListener` has a method which is left as *abstract* and which must, therefore, be over-ridden by the programmer. (This is the `actionPerformed` method). The `Action` interface is implemented by "do nothing" methods. Noting happens and the default operation of an `Action` is the same as for an `ActionListener`. But, if the programmer wishes, it is possible to over-ride the `Action` methods to, for example,

- add an icon to the GUI component
- add an accelerator key
- add a tool tip
- add a detailed description

These are desirable properties of modern user interfaces.

Interested readers could experiment with adding icons and tool tips to the `EScope` GUI as it is developed.

Text-Oriented Components

There are two component types in Swing which deal with text input: `JTextField` and `JTextArea`. Both have methods to "set" the displayed text:

```
void setText(String t)
```

and to "get" text which a user has typed in:

```
String getText()
```

Our `ComponentFrame` example used a `JTextArea`. We noted how it changed its size dynamically as a user typed text into it. In general, `JTextArea`s can be specified with a certain number of rows and columns and can be line-wrapped. By contrast, a `JTextField` is a single line of text whose default width is that of the initial displayed text (although this can be adjusted).

3.5.1 Text Fields and the Model-View-Controller Design Pattern

If you have modified your `ComponentFrame` example to use a `JTextField`, as suggested in the programming exercises, then now is the time to experiment with having two `JTextField`s linked together. Modify your program to resemble the following

```
public ComponentFrameMod()
{
// Components for the GUI
    ...
```

```
    JTextField text = new JTextField("Type here!");
                   // A text field
    JTextField text1 = new JTextField(12);
                   // A text field of width 12 'columns'
    text1.setDocument(text.getDocument());
                   //'text1' has the same 'document'
                   //object as 'text'
    ...
    componentPanel.add(text);
    componentPanel.add(text1);
    ...
}
```

Now run your modified code. You will see two text fields with the same contents, one of which is longer than the other. Now go to the shorter field and erase its contents. You will see that the other text field also becomes progressively erased as you erase each character. Now type new text into the shorter field. You will notice that this field fills up until the text window is full and the text display then begins to scroll to the left. At this juncture the longer field is still not full and you can see the new text being added to this component. Keep playing with the text fields by inputing, erasing and editing text and by scrolling around them using the left and right arrow keys. The two components provide *different views* of a single *model* object (the text string). This *separation* between appearance (*view*) and data (*model*) is used widely in Swing and is part of a design pattern known as "Model-View-Controller" (MVC). The model object for a text field is called a "document" and it can be accessed using get and set methods `getDocument` and `setDocument`. The view is the appearance of the Swing component (the little window on the text).

What are the controllers? The controllers are listener objects which listen to the component and update either its view or its model or both. Some of these controllers are the "built-in" controllers which controlled the response of our two text fields to mouse events and arrow keys. Others can be added by the programmer. You add your custom listeners to the document and they implement the `DocumentListener` interface. In particular, the `insertUpdate` and `removeUpdate` methods need to be implemented. The events triggered are of type `DocumentEvent`. You can get the text in the document by using the `getDocument` method of the `DocumentEvent` object together with the `getText` method of the `Document` object.

The MVC pattern demonstrates a key idea of design patterns in general: that of a *separation of concerns* between objects. The appearance (view) of a GUI component is a separate concern from the data (model) which it contains. Both of these are separate concerns from the behavior of the component which is governed by the controller objects. Although this design pattern is not one of the "classic" patterns to be discussed later, it can be thought of as a grouping of observer patterns (discussed in Chapter 14).

3.6 `PreEScope1`: A Simple GUI for `PreEScope0`

Appendix D.2, on page 266, shows the listing for a second revision of `PreEScope` which builds a GUI on top of the functionality of `PreEScope0`. Menu items in the main frame allow a user to

- connect to `MDSplus`
- open an experiment and shot
- request a data set (download the X and Y arrays)
- plot the data
- close the experiment
- disconnect from the server
- quit

It would be a worthy programming exercise to attempt to construct this program on your own. Regardless of whether you do or not, please play with the version of this program on the CD. Particularly consider its error handling when menu actions are requested in the wrong order. Also note that it is possible to request multiple datasets and have them displayed in different plotting windows (try `.operations:i_fault` and `.operations:diamag`). Closing one plotting window will not cause the entire program to exit because we do not specify `JFrame.EXIT_ON_CLOSE` when the plot frame is created.

When you run this version of the software, you will notice that information is requested from the user using special, pop-up dialog window. You can create these dialog windows using static methods of the `JOptionPane` class as described below. (In Section 6.6.1, we will replace these `JOptionPane` dialogs with dialogs which we will construct ourselves. But coding custom dialogs is slightly complicated and we think they are best left for a later stage in this book.)

3.6.1 Using `JOptionPane` to Request Information

`JOptionPane` dialogs can be used to pop-up simple prompts, warnings and requests for information.

For example, the "Connect to server" listener, taken the `EScopeFrame` class of our `PreEScope1` example, has an `actionPerformed` method which looks like this:

```
public void actionPerformed(ActionEvent event)
{
    String serverString = JOptionPane.showInputDialog
        (EScopeFrame.this,"Specify server and port",
         "ephebe.anu.edu.au:8000");
    try
    {
        dataSource = new MDSNetworkSource();
        System.out.println("\nConnecting to server ..");
```

```
            dataSource.connect( serverString );
            System.out.println("...connection successful");
      }
         ...
}
```

Here the call to the static method `JOptionPane.showInputDialog` will cause the dialog window to pop up. As the first argument to this method, we need to specify the "owner frame" of the dialog (in this case the `Connect-ServerAction` listener is an inner class inside `EScopeFrame`). The special Java syntax for referring to an outer class object from inside an inner class is

> `OuterClassName.this`

(You should remember this syntax because it is occasionally useful for other things as well. For example, if you have an instance field in the inner class with the same name as one in the outer class then you would need to use the `OuterClassName.this` prefix to refer to the outer-class field.)

The second argument to `showInputDialog` is the instruction or request. The third argument is a default text which will appear in the text field. Specifying a default text can show the user the format the data should be specified in (in our case, with a ":" separating the server name and port number). If the user just types `Enter` then the default information will be sent to the program and the dialog will disappear. The user can also edit this information before pressing `Enter`. In any case, the entire program will wait until either the `Enter` or `Cancel` buttons have been clicked. This waiting is known as the dialog being *modal*. In general you can define them to be modal or modeless.

`JOptionPane` is a very useful class which contains other static methods to create warning and error dialogs. It is well worth while reading about it in Sun's online documentation.

3.7 Programming Exercises

The following exercises will help readers become familiar with the parts of the Java API described in this chapter.

1. Take the `BlankFrame` example and edit it to resize the frame and position it at different places on the computer screen. Readers with some Java experience might like to experiment with obtaining the screen dimensions and positioning the frame always in the center of the screen.
2. Experiment with adding additional Swing components to `ComponentFrame`. Try changing the slider bar to be vertical. Try adding some options to the slider such as "snap to grid". Play with both `GridComponentFrame` and `BorderComponentFrame` so that you can be sure that you are able to position components accurately. Become familiar with the way that `BorderComponentFrame` automatically resizes the various regions. If you

are brave enough, experiment with other types of layouts such as "Box", "Card" and "GridBag". Some of these classes are well documented in the Java API and some also have links to appropriate pages of Sun's Java Tutorials [13].
3. After reading about the `BizarreComponentFrame` make up some suitably bizarre GUI examples of your own and get the feel for the way that Swing components relate to one another.
4. Even though we have not studied the graphics details of our `Plotter` class you could try to modify it and run `SimplePlot` again. For example, you could change the color of the line, change the color of the background and change the pen type. You could modify the data file to draw simple shapes. Perhaps several shapes could be drawn on the one panel. If you are keen or experienced, you could read some Java2D documentation and figure out how to plot some other graphical objects.
5. Modify the `SimpleExpandAction` program to keep expanding a frame up to a maximum size.
6. Write a program with an action which pops up another frame with a message on it. Give this new frame a menu item which will cause the entire program to exit.
7. Go back to the `ComponentFrame` example and try changing the `JTextArea` to a `JTextField` to follow the discussion in the text.

3.8 Further Reading

Once again, Sun's web documentation on the Java API [12] and Java Tutorial [13] are the first port of call for information about writing user interfaces in Java. We particularly acknowledge, and recommend, the books by Hunt ([14]), Horstmann ([15]) and Horstmann and Cornell ([21]). The two books by Topley [22, 23] are a comprehensive resource.

4

Waveform Graphics

Let us now take a more detailed look at Java2D graphics in order to understand the `Plotter` class which has been used in our introductory examples. We will then move on to constructing more beautiful graphs which will be worthy of `EScope`.

4.1 Java2D Graphics

Every regular Swing component is descended from the class `JComponent` and has a `Graphics` object associated with it. As described in Section 3.3, this object gets passed to the `paintComponent` method of `JComponent`. The actual plotting commands are contained inside this `paintComponent` method.

In order to draw a two dimensional graph, you need to define the coordinates of points, lines and shapes with respect to the coordinate system of the graph "canvas" (which is usually a `JPanel` object). You draw these by calling methods of the `Graphics` (or `Graphics2D`) object. These objects remember settings such as the background color, the line color and the line width. You never call `paintComponent` directly, but you can call `repaint()` which will force a redraw. (When should you call `repaint`? Perhaps try writing a few graphics programs and see whether you are getting into trouble and are not seeing things you thought that you had plotted. Most of the time you will see that the system does a good job of knowing when to repaint your graph. One situation when you should call `repaint` would be when you had modified the data arrays to be plotted and you wanted the graph to change immediately.)

The evolutionary nature of the Java means that there is sometimes some "baggage" that needs to be taken care of from earlier versions of the API. Having to cast your `Graphics` object to a `Graphics2D` object in order to use Java2D is one example of this baggage. Another example is that `Java2D` has been written to use *single-precision floating-point arithmetic* for calculating its shape coordinates and dimensions. There are two aspects of this which concern the programmer:

58   4 Waveform Graphics

1. The default floating point format in Java is *double precision*. You need to be aware of a possible truncation of precision when you plot floating point data. You should be careful to ensure that constants are single precision when required.
2. Although internal calculations are in single precision, the final plot is made up of points in integer *pixel coordinates*. If you have several data points which end up being mapped to the same pixel then you can ignore all but one of the data points.

Because it is so common for people to want to feed `double` constants and variables into Java2D, the API includes *two versions* of many of the basic graphics classes. One version takes single-precision data and the other takes double-precision data. For example, there are two classes which can be used to represent lines. Both of these classes are children of an abstract `Line2D` class and each has the same functionality:

```
Line2D.Float   linef = new Line2D.Float(1.0F,2.0F,5.0F,6.0F);
Line2D.Double  lined = new Line2D.Double(1.0,2.0,5.0,6.0);
```

As well as lines, there are many other basic shapes available in Java2D. For example: `Point2D`, `Line2D`, `Rectangle2D`, `Ellipse2D`, `Arc2D`, `CubicCurve2D`, and `QuadCurve2D`). The `Area` class enables the construction of more complicated figures by superimposing shapes in a number of ways. The `AffineTransform` class lets you rotate, translate, scale and shear shapes to produce other shapes. More information about these classes can be obtained by looking at the documentation for the `java.awt.geom` package in the Java API. (When using these classes, be particularly careful of methods and their return types. Just because the internal arithmetic is done using `float`s does not mean that methods cannot return `double`s!)

4.2 A Second Look at `Plotter`

We started to describe the internal structure of the `Plotter` class in Section 3.3. We shall now work through the complete listing in Appendix B.2, on page 244, and discuss the remaining aspects of Java2D which we need to understand it. Then we will do some work to improve on it.

4.2.1 Basic Set-up

The `paintComponent` method of the `DrawFrame` class from our `Plotter` program starts by defining some internal variables and then casting the `Graphics` object to a `Graphics2D` object:

4.2 A Second Look at Plotter

```
public void paintComponent(Graphics g)
{
    BasicStroke bs;                      // Ref to BasicStroke
    int i;                               // Loop index
    Line2D line;                         // Ref to line
    float[] solid = {12.0f,0.0f};        // Solid line style

    // Cast the Graphics object to Graphics2D
    Graphics2D g2 = (Graphics2D) g;
    ...
}
```

4.2.2 Setting Colors and Strokes

Colors are objects of the `java.awt.Color` class. System default colors are available as static constants of this class, including:

```
//Static constants of java.awt.Color
Color.black
Color.blue
Color.cyan
Color.gray
Color.green
Color.magenta
Color.orange
Color.pink
Color.red
Color.white
Color.yellow
```

You can define your own colors using

`new Color(a, b, c)`

where the arguments are `int`s between 0-255 representing the red, green and blue color components. (Although colors are usually defined in RGB space, a constructor which defines other color spaces is also available.) You can adjust the brightness and darkness of a `Color` object by using the methods `brighter` and `darker`.

The color of the next object to be drawn is specified by calling the `setColor` method of the `Graphics2D` object. In the code excerpt below, this method is called before drawing a rectangle object in order to draw a solid rectangle of (white) color which will be the background for the graph. `setColor` is then called again to set the color of the pen stroke. In order to cover the entire window, the size of the solid background rectangle is calculated using the `width` and `height` fields of the `Dimension` object returned by the `getSize()` method of the Swing component. (You will need to use these values if you wish your graphics window to be resizable. In fact, you will need to use them to calculate the transformation between data coordinates and pixel coordinates on the panel as described in Section 4.2.3).

```
// Get plot size
Dimension size = getSize();

// Set background color
g2.setColor( Color.white );
g2.fill(new Rectangle2D.Double(0,0,size.width,
                                    size.height));

// Set the Color and BasicStroke
g2.setColor(Color.black);
float strokeWidth = 1.0f;
bs = new BasicStroke( strokeWidth, BasicStroke.CAP_SQUARE,
                      BasicStroke.JOIN_MITER, 1.0f,
                      solid, 0.0f );
g2.setStroke(bs);
```

The `java.awt.BasicStroke` class is used to control the width, style (dashed, solid) and ends of lines. The two forms of the constructor which are commonly used are

```
BasicStroke(float width);
BasicStroke(float width, int cap, int join,
            float miterlimit, float[] dash, float dashPhase);
```

where

- `width` is the width of the line in *pixels*
- `cap` and `join` are integer flags specifying the types of end caps to be drawn on the lines and the type of joining between lines ("beveled", "mitered" or "rounded")
- `dash` is an array of floats whose even elements ([0], [2], ...) specify the lengths of visible segments in the line (in pixels) and whose odd segments [1], [3], ... specify the lengths of invisible segments. Most commonly this array would have two elements one of which would be equal to zero if the line is solid. In our `Plotter` program it was set at the beginning of the `DrawFrame.paintComponent` method.
- `dashPhase` is an offset, in pixels, from the start of the line to start the dashing pattern.

4.2.3 Transform Data and Plot the Line

The following excerpt from `DrawFrame.paintComponent` has four steps

1. Calculate the maximum and minimum values of both the X and Y data arrays.
2. Using these values of the X and Y data ranges, calculate the pixel/data scaling factors in the X and Y directions.

3. Construct arrays of scaled data.
4. Plot the line in pixel coordinates.

```
double xMax,yMax,xMin,yMin,deltaX,deltaY;
xMax=xVals[0];
xMin=xMax;
yMax=yVals[0];
yMin=yMax;
for (i=1;i<xVals.length; i++)
{
    if (xVals[i] > xMax) xMax = xVals[i];
    if (yVals[i] > yMax) yMax = yVals[i];
    if (xVals[i] < xMin) xMin = xVals[i];
    if (yVals[i] < yMin) yMin = yVals[i];
}
deltaX = xMax - xMin;
deltaY = yMax - yMin;

double xScale,yScale;
xScale = (size.width)/(deltaX);
yScale = (size.height)/(deltaY);

double[] xScaled = new double[xVals.length];
double[] yScaled = new double[yVals.length];

for (i=0; i<xVals.length; i++)
{
    xScaled[i] = (xVals[i]-xMin)*xScale;
    yScaled[i] = size.height - (yVals[i]-yMin)*yScale;
}

// Plot curve
for ( i = 0; i < xVals.length -1; i++ )
{
    line = new Line2D.Double(xScaled[i], yScaled[i],
                             xScaled[i+1],yScaled[i+1]);
    g2.draw(line);
}
```

In this code excerpt, the calculation of the xScaled array is straightforward. The yScaled array is more subtle because the pixel Y coordinate moves *downwards* on the panel from the top left hand corner. The rule to remember here is that you take the negative of the scaled values and add the panel height to it. To check that you have this transformation correct, note that when yVals[i] is equal to yMin then yScaled[i] is equal to size.height so that the point sits at the bottom of the panel.

In some circumstances you might want to adjust the scaling transformation. For example, you might need to have a border around the panel for axis labels as shown in Fig. 4.1.

4.3 A Fancier Plot

We will now consider how we can improve the quality of our plot by

- Drawing a graph with borders and titles
- Drawing axes and grid lines
- Filtering data

An example graph is shown in Figure 4.1. To obtain a nice graph like this one, we need to consider how we might incorporate a border around the plotted area. This border needs to accommodate the following "adornments":

- the *graph title* (above the top of the graph),
- the *axis labels* below the bottom line (x-label) and to the left (vertical y-label),
- the *tick labels* below the bottom line (x-tick-labels) and to the left (vertical y-tick-labels);

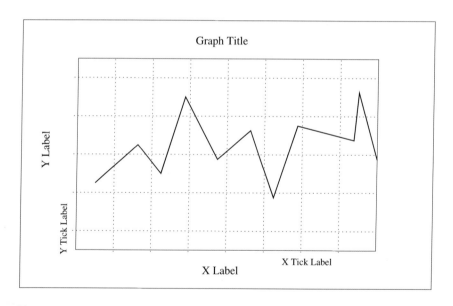

Fig. 4.1. A basic graph.

Sensible choices for the X and Y axis labels are the units of measurement for the X and Y data. `MDSplus` stores this information for each dataset. It can be obtained using the `units_of` command as follows:

```
expression=" units_of(" + address + ")";         //Y units
expression=" units_of(dim_of(" + address +"))";//X units
```

A sensible choice for the graph label would be the name of the dataset. We will assume that the axis labels and graph title have been obtained from the user of from `MDSplus` and have been sent through to the plotter class ready for use.

4.3.1 Fonts

In order to start calculating the dimensions of the border region around the plot, we will need to learn a bit about fonts in Java.

The `java.awt.Font` class handles fonts for use by the `graphics2D` class. The font is set by calling the `setFont` method, then it is drawn using `drawString`:

```
Font f = new Font("SansSerif",Font.BOLD, 18);
                   //bold font; 18 points
g2d.setFont(f);
g2d.drawString("Amps", xpos, ypos);
```

`SansSerif` is one of a number of *logical font names* which is mapped to one of a number of *physical font sets* supported by the Java run-time system. You are also able to specify the physical font sets if you want to be very exacting (but it is not altogether easy to find out what the names of the supported physical fonts actually are!). The logical `SansSerif` font is often recommended for graph headings and labels. The logical `Dialog` or `DialogInput` fonts are often used for labels on dialog boxes. `Serif` fonts can be used for other text and `Monospaced` fonts can be used if you wish to emulate a typewriter.

Fig. 4.2. A piece of text showing the meaning of the font height and ascent. The "origin" for drawing the font is shown as the solid dot on the bottom left corner of the first character.

A picture of some text is shown in Fig. 4.2. The "origin" of this text, which corresponds to the (xpos,ypos) coordinates in the call

```
g2d.drawString("EScope", xpos, ypos);
```

is shown as the solid dot to the bottom left of the first character in the text. Note that the drawing baseline is not the lower-most extent of the text and there is a block of space, known as the "descent" which is reserved for parts of characters which dip below this baseline. Similarly, there is sometimes a block of text, called the "leading", above the normal text "ascent" which is reserved for special character parts.

In order to position a block of text accurately on a plot, it is necessary to have an idea of the height and width of that text block. In order to be even more precise, it would be good to have some idea of the ratio of "ascent" to "height". It turns out that a good value for this ratio is about 0.8. We shall see how this positioning is achieved fairly accurately in several examples. In real life, you do your best to get it right and fiddle around a bit if the text is not where you would really like it to be!

The following code excerpt shows how the `getStringBounds` method of a `Font` object can be used to find the pixel width and height of a string. An argument to this method is something known as the `FontRenderContext` of the `Graphics2D` object after it has been set to the required font. (There is not much point in trying to reason about this, just make a note of it for later use!) The code excerpt also shows how a title string can be positioned in the middle of a border region above a graph. The variable `textFac` is the magic ratio of ascent to height (which we often set to be 0.8).

```
// Graph Title
g2.setFont(titleFont);   // set font
// Find size of label
FontRenderContext context = g2.getFontRenderContext();
Rectangle2D bounds = titleFont.getStringBounds(title,
                                                context);
double stringWidth = bounds.getWidth();
double stringHeight = bounds.getHeight();

int xp = (int)((xSize - stringWidth)/2.0);
int yp = (int)((borderSize - stringHeight)/2.0
               + textFac*stringHeight);
g2.drawString(title, xp, yp);
```

4.3.2 Calculating Border Dimensions

Referring back to our diagram of a basic graph in Fig. 4.1, we can see that we could center the graph on the panel by calculating the space needed for the various axis adornments and then finding the maximum of these. In fact, because the left adornments (the Y axis label and Y tick label) will probably be the same size as the bottom adornments (the X axis label and X tick label) we only need to compare the title height with the bottom adornments to calculate a reasonable border-size.

This is done in the following method, `calcBorderSize`. Although this method uses variables, such as `axisFont`, which are defined in the enclosing class, its meaning should be reasonably clear. It is included in the next version

4.3 A Fancier Plot

of our plotting class which we call `Plotter2`. It returns a border of 1.5 times the maximum combined text heights.

```
public int calcBorderSize(Graphics2D g2)
{
    /* borderSize calculation using test string "Test" */
    double border1, border2, border3, spaceFac=1.5;

    // height of top adornment: graph title
    g2.setFont(titleFont); // title font
    FontRenderContext context = g2.getFontRenderContext();
    Rectangle2D bounds = titleFont.getStringBounds
                                         ("Test", context);
    double stringHeight = bounds.getHeight();
    border1= spaceFac*stringHeight;

    // height of bottom adornments: x-axis label
    // plus x-tick-labels
    g2.setFont(axisFont); // axis font
    context = g2.getFontRenderContext();
    bounds = axisFont.getStringBounds("Test", context);
    stringHeight = bounds.getHeight();
    border2= spaceFac*stringHeight;

    g2.setFont(tickFont); // tick font
    context = g2.getFontRenderContext();
    bounds = tickFont.getStringBounds("Test", context);
    stringHeight = bounds.getHeight();
    border3= spaceFac*stringHeight;
    return (int) Math.max(border1, border2+border3);
}
```

Once the size of the blank border has been determined, our plotting program then needs to adjust its scaling calculations. It also needs to include the appropriate border offset. The following code excerpt shows how this is done (compare with Section 4.2.3):

```
    double xScale = (xSize -2*borderSize)/(deltaX);
    double yScale = (ySize -2*borderSize)/(deltaY);
    ....
        for (i = 0; i < yVals.length; i++)
        {
            xScaled[i] = (int) ((xVals[i]-xMin)*xScale
                                  + borderSize);
            yScaled[i] = (int) (ySize -borderSize -
                                  yVals[i]-yMin)*yScale);
        }
```

4.3.3 Drawing Titles and Axis Labels

Each of the title, x-label, y-label and tick values, should be drawn centered near its appropriate axis. To do this the dimensions of the label text need to be calculated from the relevant `FontRenderContext`. Careful thought needs to be given to calculating the starting coordinates of the label.

In the following code excerpt we show how to calculate the position of the small X-axis label which is to be positioned just below the bottom axis line of the graph. To do this we do the following

- start from the maximum y extent
- take off a factor multiplied by the label height; the factor is to allow some white space underneath the label
- add on a "text factor" multiplied by the label height; the text factor is roughly the ratio between the text "ascent" and the text height (see Section 4.3.1)

Now here is the excerpt. Note that the label on the X axis is `xUnits+sciXLabel`. This anticipates that scientific notation might be used to label the tick marks and that the `xUnits` label might be qualified by this suffix (for example `Amps (x10^2)`). The calculation of scientific notation will be discussed below in Sect. 4.4.

```
/* Draw X-axis label */
g2.setFont(axisFont); // set font
//find size of label
context = g2.getFontRenderContext();
bounds = axisFont.getStringBounds(xUnits+sciXLabel,
                                    context);
stringWidth = bounds.getWidth();
stringHeight = bounds.getHeight();

xp = (int)((xSize  - stringWidth)/2.0);
yp = (int)(ySize  - 1.25*stringHeight
                  + textFac*stringHeight);
g2.drawString(xUnits+sciXLabel, xp, yp);
```

4.3.4 Filtering Data

It is very important to be aware that some signals will have many thousands of data points. Most often, it will not be necessary to plot all of these points. Indeed, it is common to have many graphs displayed together on a computer screen, in which case the amount of pixel "real estate" available for an individual graph will be quite small. A most sensible optimization procedure is to filter your graph data after it has been converted to pixels and then only plot the graph points which do not overlap each other in pixel space.

There are many ways to filter data. The algorithm that we describe here has the following features

4.3 A Fancier Plot

- It makes a first pass through the data and checks to see how many X coordinates overlap after converting to pixels.
- It then decides whether it is worth filtering the data or not. (The heuristic used here is that the number of points after filtering must be less than or equal to half of the original number of points for filtering to be worthwhile.)
- If the data is to be filtered, then *two points* are stored for each filtered X pixel value: the first corresponds to the minimum Y value corresponding to that X, the second corresponds to the maximum Y value corresponding to that X. If that particular point is not filtered, then it is repeated.

The filtering algorithm starts immediately after the calculation of the xScale and yScale scaling factors. It passes through the X data values and increments a counter, npoints, whenever a new X pixel value is found:

```
double  xScale = (xSize -2*borderSize)/(deltaX);
double  yScale = (ySize -2*borderSize)/(deltaY);

/* Check how many points might be filtered out
after transforming to pixels*/
int  npoints = -1;
int  currXVal, currYVal, prevXVal = 0;
for (i = 0; i < yVals.length; i++)
{ //Find how many points can be filtered
    currXVal = (int) ((xVals[i]-xMin)*xScale
                             + borderSize);
    if (npoints == -1 || currXVal != prevXVal)
    {
        npoints++; // increment counter
                   // if x pixel value different
        prevXVal = currXVal;
    }
}
```

The next step is to set up arrays to receive the scaled data. This step anticipates that two data points will be stored for every filtered X value if filtering is to be switched on:

```
// Set up arrays to received scaled points
int  nDim = yVals.length;
if (npoints <= yVals.length/2) nDim = 2*(npoints +1);
double[] xScaled = new double[nDim];
double[] yScaled = new double[nDim];
```

The actual filtering now takes place. The first part of the if statement corresponds to the case of there not being enough points to filter and is unchanged from the previous version of Plotter:

```
if (npoints > yVals.length/2)
{//Heuristic condition for "not enough points to filter"
```

```
        for ( i = 0; i < yVals.length; i++)
        {
            xScaled[i] = (int) ((xVals[i]−xMin)*xScale
                                            + borderSize );
            yScaled[i] = (int) (ySize −borderSize
                                    − (yVals[i]−yMin)*yScale );
        }
        npoints = yVals.length;
}
```

The second block of the if statement does the work. If the points have different X pixel values then they are repeated, otherwise the minimum and maximum values of the corresponding Y pixel values are calculated and stored:

```
else
{ //It makes sense to filter. Adjacent filtered points
  //store max/min Y vals
    npoints = −1;
    for ( i = 0; i < yVals.length; i++)
    {
        currXVal = (int) ((xVals[i]−xMin)*xScale
                                        + borderSize );
        currYVal = (int) (ySize − borderSize
                                − (yVals[i]−yMin)*yScale );
        if (npoints == −1 || currXVal != prevXVal)
        { // Points have different xVals; store two copies
            npoints++;
            prevXVal = currXVal;
            xScaled[2*npoints] = currXVal;
            yScaled[2*npoints] = currYVal;
            xScaled[2*npoints + 1] = currXVal;
            yScaled[2*npoints + 1] = currYVal;
        }
        else
        { // Points have same xVals; store max/min yVals
            if (currYVal < yScaled[2*npoints])
                    yScaled[2*npoints] = currYVal;
            if (currYVal > yScaled[2*npoints+1])
                    yScaled[2*npoints+1] = currYVal;
        }
    }
    npoints = 2*npoints+1;
}
```

4.3.5 Overall Structure of paintComponent for an Adorned Graph

The overall structure for paintComponent method for an adorned graph is similar to that in Plotter. The major differences are that the size of the blank border needs to be calculated before the data points are scaled and that the axes, ticks and tick labels need to be drawn *before* the x and y axis labels are drawn. (The reason for this last point is that the axis labels may need to be modified if scientific notation is used for the tick labels.)

A complete listing of the adorned plotter program can be seen in Appendix D.3 on page 271. In summary, its `paintComponent` algorithm performs the following steps:

1. Initialize graphics context; choose colors and stroke size etc.
2. Calculate size of blank border. If the graph is to be centered, then only one border dimension needs to be calculated.
3. Perform scaling from data coordinates to pixels; possibly filter points as they are scaled.
4. Plot the curve.
5. Draw ticks, grid and tick labels for the x-Axis and the y-Axis.
6. Draw the graph title and the axis labels.

The calculation of the tick marks and labels will be discussed in the next section.

4.4 Axis Calculations: Tick Positions, Tick Values and Scientific Notation

Finding the tick positions and tick values is not as easy as it might first appear and it is the most complicated part of the, revised, `Plotter2` class shown in Appendix D.3 (page 271). It is probably best to not bother with this part of `Plotter2` on a first reading of this chapter. Readers may prefer to experiment with `PreEScope2` instead. Later, in Chapter 10, we will come back to considering the axis-drawing algorithms in detail when we refactor them using the template pattern.

The problem with simple axis-drawing algorithms is that just sub-dividing any data range into equal bins for ticks will result in tick labels which look "ugly". For example, if your X-data lay between zero and 4.26 seconds and you wanted to have 10 bins, then the bin range would be .426 seconds and the tick x-tick-labels, to 3 decimal places, would be 0, .426, .852, 1.278, and so on. It is worthwhile doing a bit of extra work to come up with "nice" bin ranges and "nice" tick labels. The algorithm described below will force these nice values to be multiples of 1,2,5 and 10.

We will also want to use *scientific notation* for our tick labels. The idea is that numbers with a fractional part can be represented as the significant digits in a *mantissa* along with the value of an *exponent*. For example:

$$250 = 2.5 \times 10^2$$
$$0.25 = 2.5 \times 10^{-1}$$

This idea is just the same as the way that floating-point numbers are stored in a computer: separate storage space is assigned to the mantissa, the exponent and the sign bit. The beauty of scientific notation is that we can choose a reasonable number of significant figures for our tick labels and then deal with data having very small or large exponents. If, for example, our time resolution happens to be in milliseconds, then we could change our axis label from "seconds" to "seconds (x10\^-3)" and then read off the x values corresponding to the data points scaled by 10^3.

So, the two issues of finding nice bin numbers and plotting points using scientific notation are linked together. We shall illustrate how they might be managed with respect to the listing for `drawXAxis` shown in Appendix D.3. This algorithm has the following steps:

- Compute the width of a dummy tick label having a decimal point and `sigFigs` digits, where `sigFigs` is the number of significant figures.
- Use the width of this dummy label to estimate the maximum number of "bins" between tick labels. The idea is that you cannot label more than a certain number of ticks without text running together.
- Calculate a bin size which has a "nice" number for its mantissa.
- Using integer division, find the first tick mark and the last tick mark.
- Find all of the interior tick marks.
- Determine whether to use scientific notation. If so, calculate the power of 10 needed to scale the maximum absolute tick value to have one digit to the left of the decimal point.
- Draw the interior tick marks and grid (if desired).
- Calculate and draw tick labels. The numbers need to be formatted appropriately after, perhaps, dividing by the relevant power of 10. Then their width needs to be calculated and the labels drawn.

To determine the nice number for the tick bin size, we first calculate a nice number for the entire data range. This nice number will be the mantissa of a real number in scientific notation and it will be either 1, 2, 5, or 10 after a form of "rounding to nearest".

1. Calculate $e(x) = \text{floor}(log_{10}(x))$ which will give the exponent value
   - Example 1: If $x = 15.3$, $log_{10}(x) = 1.18$ and $e(x) = 1$.
   - Example 2: If $x = 0.153$, $log_{10}(x) = -0.8153$ and $e(x) = -1$.
2. The mantissa is $f(x) = x/10^e$.
3. For *rounding to nearest* find a nice number for f, nf:
   - numbers less than 1.5 are rounded to 1
   - numbers between 1.5 and 3 are rounded to 2
   - numbers between 3 and 7 are rounded to 5
   - numbers greater than 7 are rounded to 10
4. Once we have this first nice number, nf, we find the nice number for the tick bin size by dividing it by the number of bins and "rounding up":
   - numbers less than or equal to 1 are rounded to 1
   - numbers between 1 and 2 are rounded to 2
   - numbers between 2 and 5 are rounded to 5

- numbers greater than 5 are rounded to 10

A method, `nicenum` which computes this algorithm is shown in the listing in Appendix D.3.

Finally, the listing the `drawXAxis` method is also given in Appendix D.3. The Y-axis method is similar to the X-axis method, except that the tick marks and labels need to be rotated and drawn correctly. So the call to `g2D.drawString`, in X, is replaced by the following commands, in Y:

```
// Rotate and draw string; rotate back
g2D.rotate(-90.0*Math.PI/180.,xp,yp);
g2D.drawString(tickLabel,xp,yp);
g2D.rotate(90.0*Math.PI/180.,xp,yp);
```

4.5 PreEScope2: Nicer Graphs from `MDSplus`

The idea of this revision of `PreEScope1` is to incorporate the fancy plotter, `Plotter2`, which we have been discussing in this chapter.

A full listing of `Plotter2` can be found in Appendix D.3. Some readers might enjoy the challenge of writing this class themselves using the information given in this chapter. We do not necessarily recommend it, but, if you do decide to try, remember that the `Plotter2` class can also be used with `SimplePlot` which will be faster for debugging.

Using either the supplied version of `PreEScope2` or your own code, you can now complete the following exercises.

4.6 Programming Exercises

The following exercises will help readers become familiar with the parts of the Java API described in this chapter.

4.6.1 Programming Exercises

1. If you have not already done so, you may wish to experiment with the `Plotter` class by changing the color of the background and the pen stroke. Try implementing dashed lines. Also try plotting arrays of data which correspond to recognizable shapes. Then try plotting an analytic function such as a sine curve.
2. Try out `PreEScope2` on a number of different data-sets such as

    ```
    .operations:diamag
    .operations:i_ovf
    .rf:i_bot
    .rf:rf_drive
    ```

3. Experiment with aspects of the `PreEScope2` plot
   - adjust the number of significant figures
   - change fonts and font-sizes
   - resize the window and notice how the labels change
   - scale the data and notice how scientific notation works
   - play with the nice-numbers algorithm for the tick labels; swap the rounding types (round to nearest and round up); choose only one rounding type

4.7 Further Reading

In addition to Sun's website, and the Java texts mentioned in previous chapters, "Java2D Graphics" by Knudsen ([24]) and "Java for Engineers and Scientists" by Chapman ([25]) are worthwhile reading on the subject of Java2D graphics. The nice-numbers algorithm is found in the chapter by Heckbert in the book Graphics Gems [26].

5
Interactive Graphics Using Mouse Events

Interactive graphics is huge! It is a major advantage of modern languages, like Java, that you are not only able to produce nice plots of your data but you are able to write code which enables an interactive probing of that data. This interactive probing is often done using the mouse.

Of course, the GUI for `PreEScope` that we have been constructing is already interactive because clicking on it can cause a menu list to drop down. This is an example of the built-in interactivity of Swing components discussed in Section 3.1.2. In this chapter, we will discuss the low-level event handling of mouse behaviour and show how it can be used to incorporate an interactive tool into our `Plotter` class.

The way you handle mouse events on graphs is very similar to constructing listener classes for Swing components described in Sec. 3.4. As discussed there, Java's Swing classes use the AWT layer of event handling. There is *one event queue* and and several event classes. Event objects get inserted into the queue as they are generated. Listener objects monitor the queue and process appropriate event objects when they find them.

The `java.awt.Component` class, which is inherited by all Swing components, has two methods to add mouse listeners. For example:

```
addMouseListener( mouseHandler );
    //'mouseHandler' is a mouse listener object
addMouseMotionListener( mouseHandler );
```

The mouse listeners provide implementations of two interfaces, `MouseListener` and `MouseMotionListener`. These contain methods to handle events such as clicking and dragging the mouse.

5.1 Mouse Interfaces and Events

There are two important interfaces which need to be implemented if we are to catch mouse events. Their methods pass `MouseEvent` objects.

5.1.1 The `MouseListener` Interface

The `MouseListener` interface contains three methods which are called whenever a user clicks a mouse button:

```
mousePressed ( MouseEvent  e )
mouseReleased ( MouseEvent  e )
mouseClicked ( MouseEvent  e )
```

It is up to the programmer to decide which of these methods will be used. To all intents and purposes there is usually little practical difference between, one the one hand, a mouse press followed by a release, and, on the other hand, a complete mouse click. But these three methods together do allow us to add extra functionality to our mouse handling.

There are two further methods in this interface:

```
mouseEntered ( MouseEvent  e )
mouseExited ( MouseEvent  e )
```

These get called when the mouse cursor enters or exits a particular GUI component and they can be used to change the shape of the cursor when it is over a particular component.

The `MouseListener` interface is implemented, as "do-nothing" methods, in the `MouseAdapter` class.

5.1.2 The `MouseMotionListener` Interface

The `MouseMotionListener` interface contains two methods which are called whenever a user clicks a mouse button:

```
mouseDragged ( MouseEvent  e )
mouseMoved ( MouseEvent  e )
```

The most useful of these is `MouseDragged`. A stream of calls to this method takes place when a user clicks the mouse on top of the relevant component and then holds the mouse button down and moves the mouse. This continues until the mouse button is released, even if the cursor moves beyond the boundary of the component. It is common for the one mouse listener class to implement both `MouseListener` and `MouseMotionListener`.

The `MouseMotionAdapter` class implements the `MouseMotionListener` interface as do-nothing methods.

5.1.3 The `MouseEvent` Class

This is an extremely useful class which provides us with the tools we need to write powerful mouse listeners. A `MouseEvent` object gets sent to the appropriate listener methods. It can be interrogated using its accessor methods:

- `public int getButton()` returns an `int` which corresponds to the particular mouse button used. This `int` corresponds to the static fields `MouseEvent.BUTTON1`, `MouseEvent.BUTTON2`, `MouseEvent.BUTTON3`, `MouseEvent.NOBUTTON`.
- `public int getClickCount()` returns the number of mouse clicks associated with this event. It is used for implementing double clicks.
- `public static String getMouseModifiersText(int modifiers)` returns a string which identifies the modifier keys and the mouse button which was down at the time of the event. The special, `int`, modifiers mask which is required as input to this method is obtained by calling the `getModifiers()` method of `InputEvent` - the parent object to `MouseEvent`. Modifier keys are often "Shift", or "Crtl".
- `getX()`, `getY()`, `getPoint()` all return the pixel coordinates of the mouse relative to the component. These can be used to select particular points, or parts, of plotted data.

5.2 `PreEscope3`: The Graph Point Diagnostic

Using what we know about mouse event handling, it is a relatively straightforward exercise to add a "graph point diagnostic" function to `Plotter2`. (It should be possible for readers to do this after reading the following discussion, but the full source is, of course, available on the CD.)

The idea of the graph point diagnostic is that dragging the mouse across a waveform will find the X and Y data values of the graph point *which has the same x-pixel coordinate as the mouse*. This diagnostic will look like a cross-hair which is "glued" to the graph and which slides over the graph as the mouse is dragged. In the next chapter, the coordinates will be sent to listeners which do something with them. (One listener might print out the data values on a label field of the GUI.)

In order to construct this diagnostic, we need to accomplish the following

1. Create a mouse-handler class which implements the `MouseListener` and `MouseMotionListener` interfaces. This class will keep track of the X-pixel and Y-pixel positions of the last dragged mouse event.
2. The outer class probably should keep a reference to the mouse handler and it needs to "add" it to the appropriate GUI component (itself).
3. In another part of the program, it will be necessary to find the data point having the closest X-pixel value to the cursor. This can be easily done if the original data have already been transformed to pixels. (If not, then

76   5 Interactive Graphics Using Mouse Events

the X-pixel value can be transformed to data coordinates and compared with the data array.)
4. The large cross-hairs need to be drawn on top of the graph corresponding to the last cursor position.
5. If desired, appropriate methods need to be written and called to notify listeners. This is not implemented in the present example. It will be deferred to `PreEScope4` in the next chapter and will later be used to illustrate the observer pattern.

The following code excerpts illustrate the first four steps. They are taken from the `DrawPanel` class of the `Plotter3` file. As usual, the source code for `PreEScope3` is available on the accompanying CD.

- An inner class is inserted into the `DrawPanel` class. In this case it implements both the `MouseListener` and the `MouseMotionListener` interfaces. Notice how a `repaint` command is issued from within the `mouseDragged` method. The program will not work properly if you comment this line out. (Try doing this!)

```java
/** inner class to handle mouse events */
private class GraphMouseHandler implements
                 MouseListener, MouseMotionListener
{
    private int currX=0, currY=0;
    public GraphMouseHandler(){}
    public void mousePressed(MouseEvent e)
    {
        currX = e.getX();
        currY = e.getY();
    }
    public void mouseMoved(MouseEvent e){}
    public void mouseClicked(MouseEvent e){}
    public void mouseEntered(MouseEvent e){}
    public void mouseExited(MouseEvent e){}
    public void mouseDragged(MouseEvent e)
    {
        currX = e.getX();
        currY = e.getY();
        DrawPanel.this.repaint();
    }
    public void mouseReleased(MouseEvent e){}
    public int getCurrX(){ return currX;}
}
```

- The outer class, `DrawPanel` contains the following instance variable:

    private GraphMouseHandler mouseHandler=
                            new GraphMouseHandler();

    and adds the handler to itself in its constructor:

5.2 PreEscope3: The Graph Point Diagnostic

```
addMouseListener ( mouseHandler );
addMouseMotionListener ( mouseHandler );
```

- The `paintComponent` method of `DrawPanel` is augmented by the following block which finds the closest data index to the cursor and draws large cross-hairs on the plot. (This block comes after the waveform is plotted.)

```
/* Draw mouse cursor */
// First find index of closest x data point
int currX = mouseHandler.getCurrX ();
if (( currX > borderSize) &&
                     ( currX < xSize−borderSize ))
{
    // find closest data value from xScaled
    int minIdx = 0;
    double minDist = Math.abs( currX − xScaled [0]);
    double currDist=minDist;
    for ( i = 1; i < npoints; i++)
    {
        currDist = Math.abs( currX − xScaled [ i ]);
        if( currDist < minDist)
        {
            minDist = currDist;
            minIdx = i;
        }
    }
    // draw large cursor marker on graph
    Line2D xLine = new Line2D.Double
                (borderSize, yScaled [ minIdx ],
                 xSize−borderSize , yScaled [ minIdx ]);
    Line2D yLine = new Line2D.Double
                (xScaled [ minIdx ], ySize−borderSize ,
                 xScaled [ minIdx ], borderSize );
    g2.draw( xLine );
    g2.draw( yLine );
}
```

5.2.1 OpenGL Hardware Acceleration

A feature of modern graphics cards is that their architecture is especially tuned for repetitive graphics calculations. In order to employ this hardware acceleration, you may need to run the Java virtual machine with the following option set:

```
java -Dsun.java2d.opengl=true PreEscope3
```

5.3 Programming Exercises

The following exercises will help readers become familiar with the parts of the Java API described in this chapter.

1. Starting with your own code or the version of `PreEscope3` from the accompanying CD, experiment with commenting out the `repaint` command. In this case the graph will be repainted only when the system says that it should be. If you resize the window it gets repainted and you will see the cross-hair where you stopped dragging the cursor. If you place another window over part of the plot and then bring the plot to foreground again, then the plot will be updated - but only the part of the plot which was obscured; you can end up with repainting only part of your cross-hair!
2. Change the dragged shape from a cross-hair to a little box.
3. Implement other mouse events. Print out the relevant pixel value when you click with the mouse.
4. Practice catching double-clicking events and events corresponding to the different mouse buttons. Assign these events to suitably interesting and bizarre manipulations of your graph!

5.4 Further Reading

Once again, Sun's web documentation on the Java API [12] and Java Tutorial [13] are recommended for further information about mouse event handling. We have followed some informal patterns due to Horstmann and Cornell ([21]) in our examples.

6
Navigating the Database

As a final step to putting all of the basic `EScope` functionality together, we need to have a way of navigating around the `MDSplus` database. This database is similar to the directory and file hierarchy on a computer. It does not, intrinsically, support relational queries and there are no guards against data redundancy or updating errors. This may seem primitive to many who work in information technology, but it is often this way with e-Science databases which have grown out of sets of accumulated scientific measurements or simulations. (It is also representative of the way that many large websites manage their data!) If we wish to improve these e-Science databases by adding structured queries and error-checking capabilities, then it is possible to achieve this by constructing relational metadata databases which index the original databases.

We would like to be able to explore an `MDSplus` database using a graphical interface. The natural graphical representation is the structured tree hierarchy of folder and file icons which you get when browsing files on a computer and this is provided by the Swing `JTree` class. It allows folders to be interactively opened and shut and files can be selected to view the data – perhaps as a waveform plot.

In order to properly understand the `JTree` class, we will now step through some aspects of tree data-structures and discuss how computer codes can be written to explore them automatically. We will do this in a, slightly-labored, pedagogical fashion which starts first with lists, then trees and then deals with the corresponding Swing classes. Some readers may wish to skip over the early parts of this chapter, but please rest assured that our pedagogy will not be wasted. We will use the list examples much later on in our implementation of the Proxy pattern in Chapter 15.

Finally, we will improve our `PreEScope` GUI to be quite sophisticated. This will be the final stage of development before we start to restructure it using design patterns in Part II of this book.

6.1 Custom-Built Linked Lists

Imagine that we wanted to "cache" the data that we are retrieving from the `MDSplus` server. For example, if we are browsing a number of graphs, we might wish to have a facility which enabled us to quickly recall a graph which we had been looking at a short while ago. Saving a "history list" of graph data would be one way of speeding up this data recall. It turns out that one of the best data-structures to use for this purpose is a type of linked list.

The `ShotData` class, shown below, could be used to construct objects to hold the dataset corresponding to a particular `MDSplus` graph. It contains the fields `shotNo` and `title` which, together, identify the data uniquely. (For example, there will only ever be one data-set which corresponds to the `.operations:ifault` file of shot number 37025 – but we could make this even more precise by including the "experiment" field as well.) Our history list will contain the usual data that we are concerned with: the X and Y axis labels and the X and Y data arrays.

```java
public class ShotData
{
    private String title, xUnits, yUnits;
    private int shotNo;
    private double[] xVals=null, yVals=null;
    public ShotData(int shot, String t, String xtit,
                    String ytit, double[] x, double[] y)
    {   shotNo = shot;
        title = t;
        xUnits = xtit;
        yUnits = ytit;
        xVals = new double[x.length];
        for (int i=0; i<x.length; i++) xVals[i] = x[i];
        yVals = new double[y.length];
        for (int i=0; i<y.length; i++) yVals[i] = y[i];
    }
    //.........(get and set methods appear here)
    public String toString()
    {   return shotNo + " " + title;
    }
}
```

In order to construct a linked list, we need to have an additional class which represents a "node" on the list. This is shown in the `ShotDataNode` listing below. This class has an interesting structure: it contains a field of the same type as the class itself. This `next` field of `ShotDataNode` is meant to "point to" the "next" `ShotDataNode` object in our linked list.

6.1 Custom-Built Linked Lists

```
public class ShotDataNode
{
    private ShotData thisData=null;
    private ShotDataNode next=null;
    public ShotDataNode(ShotData inData)
    {
        thisData = inData;
        next = null;
    }
    public ShotData getShotData() { return thisData;}
    public ShotDataNode getNext() { return next;}

    //Set next field of node to point to a ShotDataNode.
    public void setNext(ShotDataNode nextNode)
    {
        next = nextNode;
    }
    public String toString()
    {
        if (thisData!=null) {return thisData.toString();}
        else {return null;}
    }
}
```

Our linked list is nothing more than a collection of ShotDataNode objects with the first one pointing to the second which points to the third, and so on, as shown in Fig. 6.1. At the very end of the list is a node which has null in its next field. There is no limit on the number of elements in such a linked list and it is very easy to break and recombine the links between nodes. The list would be addressed using a "header" node, which is the list field of the ShotDataList class shown below. It is a very fast operation to restore the last saved data set corresponding to the top node, using the pop operation, but it is a slow operation to search through an entire list and retrieve a particular data-set, using the getShotData operation. (If a list only admits operations to *push* nodes onto the "top" node, and to *pop* them off again, it is known as a stack.)

The listing for ShotDataList is shown here.

```
public class ShotDataList
{
    private ShotDataNode list=null; // Header node
    public ShotDataList() {}
    //Add data to the top of the list
    public void pushShotData(ShotData newShotData)
    {
        ShotDataNode node = new ShotDataNode(newShotData);
        if (list == null) { list = node;}
        else
```

82   6 Navigating the Database

```
        {
            node.setNext(list);
            list = node;
        }
}
// Remove the top node and return its data
public ShotData popShotData()
{
    ShotDataNode top = list;
    list = list.getNext();
    return top.getShotData();
}
//Search for a particular data-set.
//Return the data but do not remove the node.
public ShotData getShotData(int shotNo, String title)
{
    ShotDataNode thisNode = list;
    boolean found = false;
    while ((thisNode != null) && (found == false))
    {
        if ((thisNode.getShotData().getShotNo()==shotNo)
            && (thisNode.getShotData().getTitle().
            equals(title)))
                found = true;
        if (found==false) thisNode = thisNode.getNext();
    }
    if (found==true)
    {   return thisNode.getShotData();}
    else
    {   return null;}
}
}
```

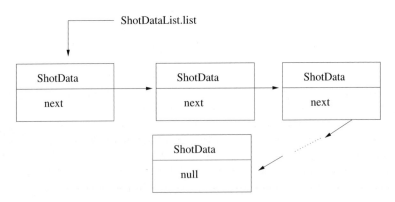

Fig. 6.1. A linked list.

There are many different ways of constructing linked-lists. For example, we could have chosen not to remove the top node in the `popShotData` method. For another example, we could have chosen to have handles on both the "header" and the "tail" node and to insert objects onto the head and remove them from the tail (a data-structure known as a *queue*). For yet another example, we could have chosen to have two links per node ("next" and "previous") which would allow you to move forwards and backwards along the list. Such a doubly-linked list is shown in Fig. 6.2.

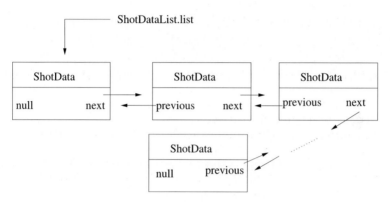

Fig. 6.2. A doubly-linked list.

6.2 Lists in Swing

The Java API contains several classes which define list-like data-structures. The most sophisticated of these are in the Java "collections framework" in the `java.util` package. These include a ready-made `LinkedList` class and there are also `Vector` and `ArrayList` classes which are a frequently-used "list emulators" although they are actually based on arrays.

There is also a list called a `DefaultListModel` in the `javax.swing` package. At the time of writing, its documentation says that it "presently defaults to `Vector`" but that there are plans to make it a "real Collections" implementation in the future. The advantage of this class is that it is understood by the Swing `JList` component. When you use it to store your data, then you can feed your data directly into `JList` and have it displayed on a GUI. By default, this `JList` display appears as a panel on the GUI with a vertical list of object names (the names returned by their `toString` methods). The `JList` contents can change dynamically as illustrated in the next section.

6.2.1 Using the DefaultListModel Class

In Section 3.5.1 we observed that many Swing components have a Model-View-Controller pattern. For example, the `JTextField` component has a "document" model object associated with it. In Swing, the "model" objects for lists implement the `ListModel` interface. It is also possible to extend the `AbstractListModel` class or to use the ready-made `DefaultListModel` class.

In the `ShotDataCache2` example in Appendix B.3, on page 246, the `JList` is created using a `DefaultListModel` made up of `ShotData` objects. The data is read in using a class, `ReadPlotData`, which is similar to our earlier `SimplePlot` example of Section 2.1 except that it returns the data as a `ShotData` object. It creates a `DefaultListModel` object and inserts the `ShotData` objects for three graphs into that list. When a user clicks on a particular list item, then the corresponding plot is launched using our `Plotter3` class described in Section 5.2. (This plotter has the graph point diagnostic implemented and you can drag the cross-hair cursor over the plots.) Once a list item is selected then it gets deleted from the list until you are left with an empty list! This program is available on the accompanying CD. It needs to be run in the parent directory of the datafiles `DoperationsCi_fault`, `DoperationsCdiamag`, `DrfCrf_drive` which contain part of the shot 37025 dataset of the `h1data` experiment.

Note the following, additional, points about this `ShotDataCache2` example:

- The `JList` constructor can take an argument of type `DefaultListModel`.
- A `ListSelectionListener` gets added to a `JList` component. This listener must implement a `valueChanged` method which can act on the specific value which has been selected.
- When list items are selected, `ListSelectionEvent`'s are generated.
- The if-block corresponding to

    ```
    if (event.getValueIsAdjusting() == false)
    ```

    is needed in the listener in order to stop multiple events being registered. (Try commenting this line out and see what happens!)
- The `getSelectedValues()` method of a `JList` returns an *array of Objects*. These get cast to `ShotData` objects in our example. Even though there is only one element in our list, a single element array is still passed by `getSelectedValues()`.

6.2.2 Using the ListModel Interface

Suppose that we had already written a custom `ShotDataList` class and we wanted to display it in a `JList`. This can be achieved by having the custom list either implement the `ListModel` interface or extend the `AbstractListModel` class. In fact, the `ListModel` interface seems to be easy to implement because it just has the following methods:

```
public interface ListModel
{
    public int getSize();
    public Object getElementAt(int i);

    public void addListDataListener(ListDataListener ear);
    public void removeListDataListener(ListDataListener ear);
}
```

Through this interface, the `JList` can get a count of the number of elements and retrieve each one of them. It can also *add itself as a listener* so that its view gets notified if the collection of elements changes. There is nothing in this interface to describe the way that elements are to be stored.

In practice, people often choose to extend the `AbstractListModel` class because there is no need to worry about maintenance of the `ListDataListener` objects. There is an example on the CD, `ShotDataCache3`, which does this by implementing the `getSize` and `getElementAt` methods in terms of the other methods from the original list class. The example has also been slightly modified from `ShotDataCache2` so that list items are not deleted after they are selected.

6.2.3 Rendering List Cell Values

It is possible to install an object in a `JList` which will be responsible for actually painting the list entry. This can be advantageous when list entries are something more sophisticated than simple strings (or simple icons which can also be displayed easily). This is discussed in Sun's Java API and Java Tutorial pages and in several good books on Swing.

6.3 Trees

Both the linked list and doubly-linked list are *one-dimensional data-structures* like arrays. In contrast, tree data structures can have more than one "next" link per node. At each node, it is possible to travel down any one of the next links. You can also have doubly-linked trees with "previous" as well as "next" links.

There are various types of tree structures and some are used for special algorithmic reasons. One of the principal reasons for selecting particular trees is that they can greatly speed the *searching for* and *insertion of* data. For example, the so-called *binary tree* shown in Fig. 6.3 is organized so that the entire branch below each "parent" node contains values which are less than the parent. The branch to the right contains values which are greater than the parent node. This means that an unambiguous decision can be made at each node to determine which branch to take to find a particular data element.

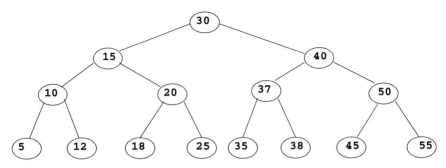

Fig. 6.3. An example of a binary search tree.

In an `MDSplus` database, the tree containing data for a particular shot is not ordered in any particular way. Instead, it resembles a directory structure on a computer. Each parent node can have any number of children. In fact, it is even more complicated because `MDSplus` databases have two types of nodes, called "members" and "children", and both need to be considered separately. This annoying issue is discussed in Section 6.5.1.

6.3.1 Recursion

Java methods can call themselves. This can be illustrated by the "common classical example" of calculating the *factorial* of a positive integer, n

$$n! = n(n-1)(n-2)...1$$

This can be done using the following iterative loop:

```
public static int factorial(int n)
{   int r = 1;
    for (int i=1; i<=n; i++) r=r*i;
    return r;
}
```

It can also be done using recursion:

```
public static int factorial(int n)
{   if (n<=0) return 1;
    else return n*factorial(n-1);
}
```

Recursion is a very elegant way of thinking about some algorithms and some recursive algorithms are very fast. However it is also possible to write *very slow* recursive algorithms and care is often needed. For example, you need to be aware that there is an overhead in making all of the method calls in the recursive factorial method.

6.3.2 Using Recursion to Probe File Structures

Here is an example of using recursion which anticipates the way that we will probe the MDSplus database. The example uses facilities of Java's File class to probe a directory structure. The method traverseTree calls itself repeatedly as the tree is traversed. For each directory, an array containing the names of the files and subdirectories is created. The array is then stepped though and counters are set depending on whether a name is just a file or a subdirectory.

You can try the following DirectoryTree program out on the sample databases provided for use with the MdsipSimulator described in Appendix A.1 on page 227. Compile and then run this program in the examples directory on the CD and supply the name of the database as an argument.

```java
import java.io.*;
public class DirectoryTree
{
    private static int noDir=0, noFiles=0;
    public static void main(String[] args)
    {
        if (args.length !=0)
        {
            traverseTree(args[0]);
            System.out.println("Number of directories is "
                                                    + noDir);
            System.out.println("Number of files is "
                                                    + noFiles);
        }
        else
        {
            System.out.println("Usage: java DirectoryTree"
                                                    + " top");
        }
    }
    public static void traverseTree(String node)
    {
        File thisDirectory = new File(node);
        String[] files = thisDirectory.list();
        for (int i=0; i<files.length; i++)
        {
            File f = new File(node + File.separator
                                                    + files[i]);
            if (f.isDirectory())
            {
                noDir++;
                traverseTree(f.getPath());
            }
            else { noFiles++;}
        }
    }
}
```

}

6.4 Trees in Swing

The Swing `JTree` component is very elaborate but it has many similarities with the `JList` discussed above. It also follows the Model View Controller pattern and we will find that there is a useful Swing class, `DefaultTreeModel` (which implements the `TreeModel` interface), which can be used to construct it:

```
TreeNode root = ........;
DefaultTreeModel model = new DefaultTreeModel( root );
```

Here `TreeNode` is another interface and the concrete object represents the root node of the tree being constructed.

Java provides a concrete implementation of `TreeNode` called `DefaultMutableTreeNode`. You can build up a tree by adding references to the children of any node from its parent. For example, the following code makes `data0` a parent node to `data1` and `data2`:

```
DefaultMutableTreeNode data0
        = new DefaultMutableTreeNode(".operations");
DefaultMutableTreeNode data1
        = new DefaultMutableTreeNode(":i_fault");
DefaultMutableutableTreeNode data2
        = new DefaultMutableTreeNode(":diamag");
..........
data0.add( data1 );
data0.add( data2 );
```

6.4.1 Tree Paths and Tree Selection Listeners

A `JTree` object stores the complete path to every node. If you click on a `JTree` node then the full path to that can be retrieved using a method `getSelectedPath`. The `getLastPathComponent()` method of `DefaultMutableTreeNode` can then be used to extract the name of the last node in the complete tree path:

```
TreePath selectionPath = tree.getSelectionPath();
DefaultMutableTreeNode selectedNode =
        (DefaultMutableTreeNode)selectionPath.
                        getLastPathComponent();
```

Tree selection listeners are quite similar to those for lists. An object which implements the `TreeSelectionListener` interface needs to have a single method:

void valueChanged(TreeSelectionEvent event)

The event objects are of type `TreeSelectionEvent`. You can use them to find out the tree path by using their `getPath()` method (applied twice! - see the listing for `EScopeFrame` in Appendix D.5.1 on page 285).

As with lists, it is possible to specify whether multiple tree nodes can be selected at the same time.

6.4.2 Tree Cell Rendering

By default the `JTree` class uses `DefaultTreeCellRenderer` objects to draw each node. `DefaultTreeCellRenderer` extends the `JLabel` class and the label contains an icon (for the branch node or leaf node) and a name. You are able to choose different icons for the nodes or leaves and make many other changes as well as described in Sun's Java API and Tutorial [12, 13].

6.5 MDSTree and MDSTreeNode

The full listings of `MDSTree` and `MDSTreeNode` can be seen in Section D.4 of Appendix D. These classes are used in all versions of `PreEScope` and `EScope` from `PreEScope4`. Of most interest is the recursive method, `getSubTree` which is used to descend through the `MDSPlus` file hierarchy:

```
/** Part of the recursion algorithm used to create the tree
@param path The path to the parent node of the sub-tree
@param nodeName The name of this node
@param type The node type
@return The newly-created parent node of this sub-tree
@throws IOException From call to getSubTreeList
*/
private MDSTreeNode getSubTree(String path, String nodeName,
            int type) throws IOException
{
    MDSTreeNode out;
    String nodelist[];
    int i;
    out = new MDSTreeNode( nodeName, type );
    // Process MDSplus "member" nodes
    nodelist = getSubTreeList( path + ":*" );
    if ( nodelist != null )
    {
        for ( i = 0; i < nodelist.length; i ++ )
        {
            out.add(getSubTree(path + ":" + nodelist[i],
                    nodelist[i], MDSTreeNode.TYPE_MEMBER ) );
        }
    }
    // Process MDSplus "child" nodes
```

```
        nodelist = getSubTreeList( path + ".*" );
        if ( nodelist != null )
        {
            for ( i = 0; i < nodelist.length; i ++ )
            {
                out.add(getSubTree(path + "." + nodelist[i],
                        nodelist[i], MDSTreeNode.TYPE_CHILD ) );
            }
        }
        //System.out.println(out); //Uncomment for node names.
        return out;
}
```

This method has a commented print statement near the end. If you uncomment it and run the `ExperimentTree.java` program available on the CD (in the `preescope4` directory) then you will obtain the names of all of the datasets as they are processed and you can see how the tree is traversed.

The `getSubTree` method calls `getSubTreeList` to get a list of the names of the child nodes of a particular parent. It does this using the `getnci` function of `MDSplus` which will be discussed in Section 6.5.1:

```
/**
Gets a list of all member and child nodes of a node
@param path The path to the parent node
@return A list of the member and child nodes
@throws IOException If there is an error communicating
 with the MDSPlus server
*/
private String[] getSubTreeList(String path)
                                           throws IOException
{
    MDSDescriptor result;
    int nodes;
    int i;
    String out[];
    String expr;

    expr = "GETNCI(\"" + path + "\",\"NODE_NAME\")";
    result = database.evaluate( expr );
    if ( result.getDtype() == MDSDescriptor.DTYPE_CSTRING )
    {
        if (result.getCstringData().endsWith(
                                        "Node Not Found" ))
            { //There are no nodes below this path
            return null;
        }
        if ( result.getCstringData().substring(0,4)
                                           .equals("Tree" ))
            { //Some other sort of error: throw exception
```

```
                throw new IOException( "GETNCI Error: " +
                                result.getCstringData() );
            }
            nodes = result.getCstringData().length()/12;
            out = new String[nodes];
            for ( i = 0; i < nodes; i ++ )
            {
                out[i] = result.getCstringData().
                            substring(12*i,12*(i+1)).trim();
            }
        }
        else
        { // Didn't get correct format from message
            throw new IOException( "GETNCI Error: " +
                            "Incorrect format returned: " +
                            result.getDtype() );
        }
        return out;
    }
```

6.5.1 Reading the MDSPlus Experiment Hierarchy

As discussed above, `MDSplus` contains two types of nodes known as "members" and "children". The `getSubTree` listing in the previous Section shows how each of these node types needs to be considered separately when traversing the directory tree. In this section, we will discuss how we can find out the location of these nodes using the `MDSplus` function `getnci`. But, before we do so, here are five reasons why the `MDSplus` tradition of having these two types of nodes, is confusing:

1. You need to put extra loops into your code to deal with each file type. See, for example, the `getSubTree` listing in the previous Section.
2. The distinction between the two node types is not very clear. A "member" node can contain data but that a "child" node cannot. So a "member" node appears to be a leaf node for the tree except for the fact that it can have descendents. The developers of `MDSplus` are aware of this confusion, but there are already many important experimental databases which have been constructed using this convention. Any future revision of the nomenclature will need to be backwards compatible.
3. The names are confusing. Both node types can be children and parents.
4. The nodes are associated with file designators which have a ":" in front of them if they are "members" or a "." in front of them if they are children. These symbols are hard to remember. (Although you can try to remember by thinking of the colon, ":", as having one dot for the next node and one dot for the data!)

5. Dots and colons are terrible symbols to use on some file systems. In particular, Unix gets confused by them. In some examples in this book, we have modified the names of data files to get rid of these symbols.

After struggling for some years to explain `MDSplus` node-naming conventions to students, we now advise students to ignore the two node types as much as possible. If necessary, write your software assuming that there is only one node type and then add in extra loops using some of our example codes as a guide. Let us move on!

If you look back to our `SimplePlotServer` example in Section 2.4.1, you will see that we constructed a simple server which responded to queries by sending data over the internet. In an analogous way, a program which connects to a remote `MDSPlus` database must ask for the names of directories and files if it needs to descend through an entire tree of the `MDSplus` database.

In order to be able to find the particular data of interest in the `MDSPlus` system, it is necessary to know where that data is stored. The `MDSPlus` function, `getnci`, can be used to obtain the file hierarchy of the experiment. It has two arguments. The first is related to the path of a parent node. When the second argument is the string "`NODE_NAME`" then the function returns a string representing a concatenated list of node names corresponding to the "members" or the "children" of that parent.

For example, if you called `getnci` as follows:

```
getnci("\\H1DATA::TOP.*","NODE_NAME")
```

then you would receive an `MDSDescriptor` object which contained a string corresponding to the "child" nodes below the root node, called "TOP", of the "H1DATA" experiment. This string might look like:

```
"ELECTR_DENS FIBRE       FLUCTUATIONSLOG       MOSS    ..."
```

In `MDSplus`, node names are restricted to a maximum of 12 characters, so each node name can be obtained using the `substring` and `trim` methods of the `String` class. In this case the "child" nodes of the top level of the hierarchy are listed. To obtain the "member" nodes of the top level the following expression would be used with a colon before the "*":

```
getnci("\\H1DATA::TOP:*","NODE_NAME")
```

To obtain the whole hierarchy, recursion can be used to continually list all the "member" and "child" nodes for a particular node. If a node has no "member" or "child" nodes then an error string is obtained which *ends with*:

```
Node Not Found
```

An example of the operation of `getnci` to traverse a tree can be seen in the listing of `getSubTreeList` in the previous section.

6.6 PreEScope4: A Waveform Browser for MDSplus

Now we come to the final stage of the "PreEScope" phase of our case study. The idea of this revision, PreEScope4, is to take PreEScope3 and weave it together with, firstly, the classes already provided in this chapter to traverse the MDSplus database and, secondly, some additional hints which we provide here. The goal is to end up with a GUI which looks something like Fig. 6.6. This GUI has a drop-down menu ("File") to connect to the server and open an experiment. It has a panel which will display the tree structure of the MDSplus database. When a file containing data is clicked on, the data gets plotted in the graphics panel on the right hand side. Above this panel is a JLabel which is used to print information such as the path to the selected file or the coordinates of the cross-hair cursor on the waveform plot. Readers are encouraged to try this phase of development themselves after reading the rest of this section. But do expect it to be a reasonable amount of work. Most of the classes you need for the solution are provided in Appendix D.5 and the entire code is on the CD.

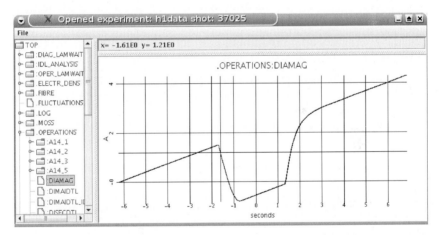

Fig. 6.4. Graphical User Interface for PreEScope4.

We consider that the specifications for this version of the software have been informally specified by the preceding discussion. We now consider some specific issues which will sharpen this specification and provide some "tips" about constructing the software if you do decide to construct it yourself.

6.6.1 Issues to Consider for PreEScope4

There are several issues to consider when constructing PreEScope4. Many of these will arise naturally if you go about this exercise yourself.

Construction of the Main Window

The main window will have a menu which provides the following menu items:

Connect Display a dialog in which the operator can enter a server address and port number and select an OK or Cancel button. If OK is selected then an attempt will be made to connect to the specified MDSPlus server.

Open Display a dialog in which the operator can enter an experiment name and shot number and select an OK or Cancel button. If OK is selected then an attempt will be made to open the specified experiment. If successful, the hierarchy will be displayed in the JTree in the application's frame.

Exit Quits the application.

Looking at Fig. 6.6, the main window will have a "split-pane" view of the currently opened experiment. This will contain:

- A JTree which displays the hierarchy of the experiment,
- A JLabel which initially displays complete path of the currently-selected node. After a user starts to drag a cursor across a waveform plot, the coordinates will be printed on this JLabel field.
- A JPanel in which we will draw a graph for the selected node in the hierarchy.

Swing contains a JSplitPane class. A class with two split panes can be constructed by combining JSplitPanes together as in the following example:

```
rightPane = new JSplitPane( JSplitPane.VERTICAL_SPLIT,
                new JScrollPane( description ), graph );

mainPane = new JSplitPane( JSplitPane.HORIZONTAL_SPLIT,
                new JScrollPane( tree ), rightPane );
```

Dialogs

PreEScope4 will use dialogs to solicit input and output from the user. We have already used the JOptionPane class to solicit input from a user in Section 3.6.1. JOptionPane's are useful for small, single-line communication with a user. (They are not the only type of ready-made dialog available in Swing. For example, the JFileChooser dialog provides a simple mechanism for enabling a user to choose a file.)

For a much more general type of dialog, you need to use JDialog objects which are like little mini-GUIs. When JDialog frames are created they must have a parent frame. They call their setVisible method to display themselves. Rather than discourse at length about dialogs we recommend that readers consult Sun's online API documentation which contains a link to the Sun Java Tutorial pages on this subject. Alternatively, just use the connect dialog listed in Appendix D.5.2 (and modify it as appropriate to construct an open dialog). A picture of PreEScope4 with the connect dialog open is shown in Fig. 6.6.1.

6.6 PreEScope4: A Waveform Browser for MDSplus

Fig. 6.5. Connect dialog for PreEScope4.

Selecting a Node in the JTree

You can add a `TreeSelectionListener` to the tree object using

tree.addTreeSelectionListener(**new** MDSTreeSelectionListener)

and you need to define your `MDSTreeSelectionListener` class to:

- Get the full path of the selected node
- Display this text in the `JLabel` object which describes the currently selected node

Getting the path right requires some fiddling around. The following listing of the private listener class, `DataTreeSelectionListener` does the job:

```
private class DataTreeSelectionListener
                   implements TreeSelectionListener
{
    public void valueChanged( TreeSelectionEvent e )
    {    // Get new data and plot it
        Object path[];
        String stringPath="";
        path = e.getPath().getPath();
        shortFileAddress="";
        for ( int i = 1; i < path.length; i ++ )
        {
            shortFileAddress = shortFileAddress
                            + path[i].toString();
        }
        stringPath = stringPath + shortFileAddress;
        if (!shortFileAddress.equals(""))
                // (Do not plot or display the "TOP" node)
        {
            EScopeFrame.this.setDescription(stringPath);
            isGraphData = false;
            isDataError = false;
            if (EScopeFrame.this.hasSignal(stringPath))
```

```
            EScopeFrame.this.getPlotData();
            EScopeFrame.this.graph.setGraphData(xVals,yVals,
                 xUnits,yUnits,shortFileAddress,
                 isGraphData,isDataError);
            EScopeFrame.this.graph.repaint();
                 //(prints a message if no data)
```

Nice Formating of Data from the Graph Point Diagnostic

The `java.text.DecimalFormat` class provides specialized formatting for numerical data. It can be used to print out the graph point diagnostic of the graph. The following code excerpt prints out such a number in exponential notation with two places after the decimal point:

```
DecimalFormat decForm = new DecimalFormat("0.##E0");
String xString = decForm.format(x);
String yString = decForm.format(y);
description.setText(" x= " + xString + " y= " + yString);
```

Does the Node Contain Valid Data?

There are several possible error conditions which may arise due to the nature of the data being downloaded from `MDSplus`. For example, if a particular file on the tree contains signal data then there is usually enough of it to plot a graph. However, for some signals only the y-axis data is defined for a particular signal. Also, some signals do not have units defined for one of both of the axes. Finally, some signals are have different vector lengths for the x and y axis data. All of these situations need to be taken care of.

You can retrieve signal data using a sequence of commands such as the following (for the `.operations:i_fault` data-set):

- `.operations:i_fault` to obtain the y-axis values.
- `units_of(.operations:i_fault)` to obtain the y-axis units.
- `dim_of(.operations:i_fault)` to obtain the x-axis values.
- `units_of(dim_of(.operations:i_fault))` to obtain the x-axis units.

If any of the above expressions return a description containing a String such as "*TreeNNF : Node Not Found*" or "*TreeNODATA : No data available for this node*" then this is an indication that there is no x-axis or y-axis data for this node.

Similarly, if there are no units defined for a particular data set then *a string containing a single space (i.e. " ")* will be returned.

If there is no y-axis data then a graph cannot be drawn. However, if there is y-axis data but no x-axis data then x-axis data can be created that is a vector containing [1 ... *no. of elements in y-axis data*].

If both x-axis and y-axis data exist, then you should check that the length of the two arrays are the same. If they are not, then the simplest thing to do is to truncate the longer of the two to the length of the other.

Error Checking

Additional error checking needs to be considered. Here are some hints about error checking:

- What happens if a user types in illegal data for the server address, experiment or shot number?
- Is a user kept informed of which experiment/shot has been opened?
- Does your GUI open up trees which contain no data?

6.7 Further Reading

As remarked earlier, Sun's web documentation on the Java API [12] and Java Tutorial [13] are an excellent resource for the material covered in this chapter. We also acknowledge, and recommend, the books by Hunt ([14]), Horstmann ([15]) and Horstmann and Cornell ([21, 16]).

Part II

Refactoring EScope with Design Patterns

7
Object-Oriented Analysis and Design

This book assumes that you, our readers, will have experienced, or taught yourselves, the equivalent of an introductory programming course in Java. It is possible that you might have considerable experience in "procedural" programming languages like Fortran or C. If you have been keeping pace with the treatment in this book, then you might be increasing your confidence in building an object-oriented software system. But you might also be wondering how people came to design and construct some of the classes that have been described in previous chapters. You might be wondering whether some of these classes could have been constructed in a better way. Musings of this nature take us into the realms of object-oriented analysis and design. This is a vast discipline and it deserves respect.

7.1 Phases of Software Development

The essence of OO analysis and design is to discover a set of classes which encapsulate data together with methods which act on that data. The process of finding these classes, and linking them together to form software systems, emphasizes a broad, "horizontal", view of software with data at center-stage. In contrast, the top-down/bottom-up design processes associated with procedural languages emphasize a sequence of operations which will solve particular problems. Although these procedural designs often involve a transformation of data, the place of data is secondary in the overall design method.

Software development traditionally involves a number of phases having names such as "requirements", "analysis", "design", "implementation" and "deployment". The *requirements phase* gathers specifications for the software in a number of forms. There can be one or more sets of documents and these may range from informal descriptions to formal, legal contracts. Requirements documents can include specifications of pathways through an executing program and these are often formalized as "use cases" which enumerate the various "actors" who will use the software together with the steps taken when they use it. The use cases specify sequences of operations which must ultimately be supported by the software.

The *analysis phase* of software development involves the discovery of classes and the specification of the data and processing associated with those classes. The requirements documentation is analyzed for concepts which might be used for classes. Sometimes these concepts are embodied by nouns and noun-phrases in the text. Useful concepts often have non-trivial data associated with them as class attributes. The processing associated with those concepts is sometimes found by examination of verbs, verb-phrases and use case steps from the requirements documents. The outputs of the analysis phase are analysis models which are often represented by UML diagrams discussed below. But these models are meant to be applications-focussed and should *not* be a detailed software design.

For example, a requirements specification which read

> "The software will enable a connection to be made with an MDSPlus database. It will allow an experiment to be opened and an index of available data-sets to be retrieved. Individual data-sets can be plotted as wave-forms."

might result in the analysis classes shown in Fig. 7.1. In this diagram, the boundary of the system is shown as the two, vertical dashed lines. There are two classes which need to be considered, one to provide the interaction with the MDSplus database and one to encapsulate and plot the waveform data.

The analysis phase often overlaps with, and feeds back into, the requirements phase. There is ambiguity in how "abstract" the analysis phase should be in comparison with the upcoming design phase.

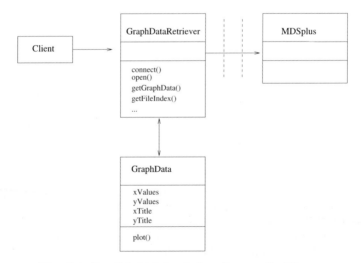

Fig. 7.1. Possible high-level class diagram for EScope.

The *design phase* takes the analysis models and produces a set of models (represented, once again, as UML diagrams) which closely mirror the classes, processing and collaborations in the target implementation language. The design phase often overlaps with the previous, analysis, phase as well as the next, implementation, phase. The most important UML diagram to be produced is the "class diagram" (or "information model"). This diagram describes the static structure of the software system. The class diagrams which we show in this book are closer to being "design" rather than "analysis".

As an example of design, suppose that we were asked to supply a subsystem of classes which would handle the interaction with MDSplus and we knew that

1. This subsystem needed to support activities to connect to a server, to open a data-set, to return an index to that data-set, to download data for a waveform plot, to close a data-set, to disconnect from a server.
2. MDSplus required data transfer to be done via MDSDescriptor objects.
3. MDSDescriptor objects needed to be binary encoded (or decoded) when writing (or reading) to MDSplus. This encoding needed to conform to a protocol specified by MDSplus.

The main data components of this problem are

- MDSDescriptor objects,
- binary decoded MDSDescriptor data, and
- the information needed to plot a waveform graph.

We would expect to discover three classes corresponding to these types. Their main processing activities would be

- supporting the interface operations for the external system (connect, open and so on),
- constructing MDSDescriptor objects corresponding to MDSplus commands and encoding these and sending them to the server,
- receiving encoded MDSDescriptor objects from the database and decoding these, and
- requesting the MDSDescriptor objects needed to construct a graph and storing the corresponding data (in a GraphData object.

The external "interfacing" requirement might lead us to construct an additional class, bringing the total to four classes. One of many subsystem designs which could satisfy these requirements might look like that in Fig. 7.2.

It is often the case that the design phase of software development will have loop-backs to the analysis phase. The next, implementation, phase can also be started before the design phase is complete. In the, so-called Unified Process [27], an important milestone of the development process is when an appropriate "architecture" for the software superstructure is obtained. This skeletal view of the software could be finalized at some time in the design phase and then remain stable through subsequent iterations between the various phases.

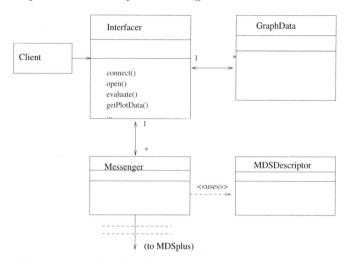

Fig. 7.2. Possible class diagram for the data server domain.

Many programmers have found that object-oriented analysis and design can be richly rewarding and that their skills in all of the development phases can be enhanced from project to project. Less experienced programmers need to have some scaffolding, and a certain "just do it" attitude to get started. One of our original ideas with this book was that design patterns could provide some of this scaffolding.

7.2 UML and Design Patterns

The subject of design patterns in object-oriented software is is becoming very influential. They can be thought of as *"...simple and elegant solutions to specific problems in object-oriented design"* [28]. The idea is that a suitably expressed design solution may be able to be applied, refined and adapted to many different problem environments. Design patterns originated in the disciplines of architecture and town planning before migrating to software. They are a sign of the maturity, but also of the difficulty, of writing good, object-oriented software systems!

In the mid 1990s, a group of computer scientists collaborated on a book "Design Patterns - Elements of Reusable Software" [28] which is still regarded as the seminal work on this subject. It is often known as the "Gang of Four" (GoF) book and it describes 23 patterns which are grouped into three categories

- creational
- structural
- behavioral

In spite of its popularity, the GoF book is definitely not an easy read! The authors have been very precise about the patterns that they describe. For each of the patterns, they consider a number of motivations and possible settings where the pattern might occur. They provide Unified Modelling Language (UML) diagrams for the pattern and then describe its implementation, advantages and disadvantages. The main languages considered are C++ and Smalltalk.

Many subsequent design patterns books have attempted to explain the GoF patterns using different implementation languages. There have been a number of "Design Patterns in Java" books and a web search on "Java" and "design patterns" will uncover a number of interesting sites (but little specifically to do with e-Science).

7.3 Design Patterns: Our Approach

In subsequent chapters we will describe the *main ideas* behind some important design patterns. We will look at a selection of some the 23 GoF patterns plus some other patterns to do with threading. Our approach is to suggest that design patterns can be thought of as targets for code *"refactoring"* (or "restructuring") as software projects grow and evolve.

We think of it this way: the basic idea of an object-oriented *language* is to encapsulate functionality and data inside a *class*. The idea of object-oriented *patterns* is to encapsulate functionality and data inside a *collection of classes*. In the same way that a well-written class might have a good chance of being reused in the future, a well-written collection of classes should have a good chance of being reused – and patterns can suggest ways in which subsystems can be reused. Our study of design patterns will be illustrated by the `EScope` case study. We will start with `PreEScope4` and we will consider how we might refactor this software using patterns.

7.4 A Diagrammatic Notation: "sUML"

Prior to the Unified Modelling Language, there were several schools of thought about modelling object-oriented software. UML brought some of these schools of thought together. Although the word "unified" conveys ideas of consistency, in UML it turns out to be closer to the mathematical idea of "union". UML is an aggregation of different diagrammatic conventions and it can be quite frustrating to sort through them all.

The approach taken in this book is to pretend to define a "small UML" subset of notation which we will use for our own purposes. Our sUML class diagrams will be as close to UML as we can make them and we will try not to include notation that is outside of UML. But we will not need to worry about offending UML purists with some of our conventions because, after all, we are using sUML notation rather than UML!

UML (and sUML) class diagrams are collections of rectangular boxes with lines drawn between them. The boxes are meant to represent classes and interfaces. The lines are meant to represent *relationships* between classes and interfaces. The idea is that the diagrams should display the connectivity between components of a software system in much the same way that a circuit diagram displays the connectivity between electronic components. UML class diagrams are often used in industry but rarely in scientific software development.

UML shows classes as boxes with three compartments. The top compartment contains the class name and can also contain a special label if it is an abstract class or an interface. The middle compartment contains the names of attributes. The lower compartment contains the names of (non-trivial) methods. Methods can be annotated to show whether they are public or private. We will follow all these conventions in our sUML class diagrams. We would like to know whether the box was a class, abstract-class or interface and we would like to know its name. Apart from these things, we will be relaxed about how much information is displayed.

7.4.1 Associations

An association is a "structural relationship that specifies that objects of one thing are connected to objects of another" [29].

What does "structural" mean? You can think of it as meaning that one (or both) of the objects will contain a permanent reference to the other object. If both of the objects have links, then the association is called "two-way". Associations are represented as solid lines between boxes representing classes and interfaces. If the association is one-way, then the line can be an *open arrow* with the arrow-head pointing to the class or interface which is referenced. We will try to remember to represent two-way associations by a line with an open arrow on each end, but it is a common convention that a line without arrows can represent a two-way association.

Associations are usually coded as *class data attributes*. If one class contains an attribute which has the type of another class then there will be an association between those two class types and the "arrow" on the class diagram will point from the containing class to the class being contained.

Note that a Java `Interface` cannot have attributes, so there should never be an arrow pointing away from a box representing a Java interface. (But a Java abstract class can have attributes.)

7.4.2 Association Multiplicities

Objects of one class may be associated with zero, one, a number greater than one, or an unspecified number of objects of another class. UML class diagrams can be annotated with these multiplicities.

Multiplicity uses the following rule: *The number marking the link attached to object A, nA, is the number of A objects that can relate to a single B object.* This is illustrated in Fig. 7.3. By convention an asterisk, "*", can represent an

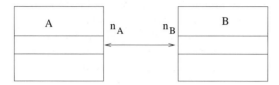

Fig. 7.3. UML showing multiplicity indicators.

indeterminate multiplicity. A range can be represented by "..". For example, "0..2" means one of 0, 1, or 2. For another example, "1..*" means greater than or equal to 1.

If the multiplicity is 1, the link between the classes will probably be implemented as single class attributes. If the multiplicity is greater than 1, then the link might be coded as a collection of attributes. For example, suppose that we wished to modify `PreEScope4` so that the main frame, `EScopeFrame`, displayed multiple `GraphPanel` graphs with two-way links between the frame and the graph panel objects as shown in Fig. 7.4. The link between the two

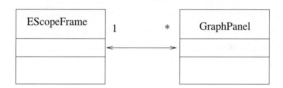

Fig. 7.4. Class diagram showing links between a main frame and multiple graph panels.

classes could be coded as:

```
public class GraphPanel extends JPanel
{
    private EScopeFrame gui;
    ...
}
```

```
public class EScopeFrame extends JFrame
{
    private ArrayList<GraphPanel> graphPanels;
    ...
}
```

If the arrow on Fig. 7.4 were one-sided from `EScopeFrame` towards `GraphPanel`, then only the `EScopeFrame` class would contain corresponding attributes of the `GraphPanel` type. (The `gui` attribute in `GraphPanel` would disappear.)

There is no need to explicitly list associations in the attribute compartment of a UML class box because you can see them by following the link.

7.4.3 Association Labels

Some people like to annotate associations with a label that describes the relationship. Others prefer to number the associations and then to label each side of the association with words describing the role played by objects of that class. You will find it tedious to keep putting in labels on associations. If you find yourself repeatedly writing "uses" or "has", then leave the labels out!

Here are some rules for using verbs or verb phrases to label associations. We shall adopt these rules in our sUML:

- If the verb (phrase) is in the *active* voice, then place it next to the active subject.
- If the verb (phrase) is in the *passive* voice, then place it next to the passive subject.
- Use a filled triangle to point to the class box which is the subject for the verb (phrase). The triangle should sit above the verb as illustrated in Fig. 7.5. (The class name next to the verb is read first followed by the verb and then the class name on the other side of the association.)
- Only label one side of the association if you think that the meaning of the relationship is clear enough.

Fig. 7.5. An illustration of our labelling convention for associations.

7.4.4 Reflexive Associations

Binary associations can exist between objects of the *same type* and are shown by connecting both ends of an association to the one class identifier. These are called reflexive associations.

7.4.5 Ignore Aggregation!

Perhaps the most radical thing we do is to ignore "aggregation". The idea of aggregation is that some objects of one class are "made up of" objects of another class. This can be important for languages which need to explicitly manage memory because the component objects must be deleted when the composite object gets deleted. But this is not a concern with Java because of its automatic garbage collection.

Representing aggregation can be frustrating for students and it can, pretty well, be replaced by an association with a multiplicity indicator. People who wish to denote aggregation mark their links with open and closed diamonds next to the aggregating class. One useful aspect of this special notation is that inner classes can be represented by the closed diamond linkage. But, overall, we feel that the subtleties involved in representing aggregation outweigh its benefits in this book.

7.4.6 Dependency

The idea of dependency (or "uses") is represented in UML by a dotted association link (usually with an open arrow to show direction). Dependency implies that altering the details of one class might affect the operation of the other.

What is the difference between association and dependency? If one class is associated with another then altering the details of one might very well affect the other. Seen this way, association is a special case of dependency. But a dependency can occur without having a permanent structural link. For example, a method of one class may accept or return objects of another class as arguments or return types. For another example, fleeting object linkages may occur as local variables in methods of a particular class. For yet another example, static methods of one class can be invoked from an object of another class.

In fact, what people generally want to do by showing dependency on their UML class diagrams, is to indicate one of a number of *particular types of dependencies* which are *not associations*. The type of each dependency is given as a label for that dependency link. The labels are placed between double brackets of less-than and greater-than symbols. In our patterns examples in the rest of this book, we will sometimes be interested in a dependency where an object of one class *creates* an object of another. The label for this link will be <<creates>>. In situations where we think it is useful to indicate a weaker "uses" type of dependency, we will use the label <<uses>> (but these will, hopefully, be few and far between).

If your class contains methods which are passed handles on other objects as arguments, but there is no permanent link to those other classes, then the relationship is one of dependency rather than association. It can become tedious to enumerate all of the dependencies in a software system and your class diagram can begin to look like spaghetti! A better representation of these dependencies would be to construct UML "collaboration diagrams".

7.4.7 Package Associations

It is useful to be able to indicate in an abstract way that one subsystem (or "domain" or "package") is dependent on another without showing the individual classes which participate in that association. People using UML sometimes do this by drawing a dotted line, and open arrow, between the two package symbols as in Fig. 7.6. Because this is an abstract representation of an association, we will not always distinguish between associations, dependencies and inheritance when it comes to packages.

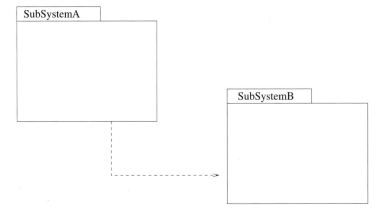

Fig. 7.6. An example of a package association.

7.4.8 Inheritance and Implementation

An inheritance relationship is shown by a *solid* line with an open triangular arrow on the end of the line next to the parent class. An "implements" relationship is shown by a *dashed* line with an open triangular arrow on the end of the line next to the interface which is being implemented. The idea of "implements" is similar to inheritance: an interface specifies the signatures of the methods in a class and an implementing class "specializes" these methods to have a particular implementation. An abstract class can be both inherited and implemented so we will mark them in the same way that we mark inheritance.

Examples of inheritance and implementation are shown in Fig. 7.7.

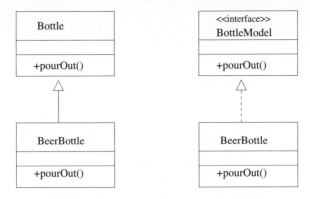

Fig. 7.7. Examples of inheritance and implementation.

7.5 Summary of Our sUML Class Diagrams

Hopefully you will get the hang of sUML class diagrams as you follow our patterns examples. A summary of their main features are

1. Boxes to represent classes, abstract classes and interfaces.
2. The boxes have three compartments
   - A *name* together with a special indicator if the box represents an abstract class or an interface. (These indicators are <> for an abstract class and <<interface>> for an interface.)
   - Class data attributes (sometimes, completely or partially, filled in)
   - Public class methods (sometimes, completely or partially, filled in)
3. Associations are represented as solid links between classes. Open arrows are used for one-directional links. Arrows on each end, or no arrows at all, for two-way links.
4. Associations are labeled with their multiplicity according to conventions described above.
5. Associations can be labeled with their role according to conventions described above.
6. Creational dependencies are shown with dashed, open arrows and the label "<<creates>>".
7. Other dependencies are sometimes shown.
8. Inheritance is shown by a solid line with open triangular arrow. Implementation is shown by a dashed line with open triangular arrow. Abstract classes are treated the same way as inheritance.
9. Aggregation is not represented.
10. Expect us to ignore our own rules from time to time! In particular, we may find it expedient in this book to draw diagrams which are illustrative rather than detailed. These diagrams might not contain all of the detail you would expect.

Remember that our sUML is not really very different from standard UML. So if you learn it, you should be in good shape to use and understand UML. We will use sUML to explain the *main ideas* behind the implementation of each pattern and to show the main features of each revision of `EScope`. As such, we can do without some of the details if they are not needed. We will not represent aggregation and will not display all of the dependency linkages. We will not always show inner classes. (We will hardly mention "sUML" in the remainder of this book, preferring to call our diagrams simply "class diagrams" or "state transition diagrams" as appropriate.)

In other situations, you might need to specify as much detail as possible in your UML (or sUML) diagrams. This might be true if you were passing on details of a software design to someone else to implement.

7.6 Further Reading

The material presented in this chapter, and all of the following chapters on design patterns, can be supplemented by reference to some of the many books on patterns and software development.

As mentioned repeatedly in this book, the seminal work on design patterns is the GoF reference [28]. Other classic references on object-oriented software development are [27, 29, 30, 31]. Hunt's book on a "Guide to the Unified Process featuring UML, Java and Design Patterns" [32] is a readable account of these four subjects. Scott's guide to UML, "UML Explained" [33], is clear, concise and practical. Mellor and Balcer [34] describe a project which adds semantics to UML and makes its models executable and testable. There are several other books and web references on object-oriented design patterns using Java and other languages.

8
First Facades

Consider a possible description of `PreEScope4` (see Section 6.6):

> "When the program starts up, a GUI window appears. The window has 4 parts:
> - a menu bar with a drop-down menu called `File`,
> - a panel which can display a directory tree,
> - a panel which displays text – either the path to a selected file or the coordinates of the graph point diagnostic on a graph,
> - a panel which displays the waveform plot of a selected dataset.
>
> The `File` menu has menu items for `Connect`, `Open` and `Exit`. The `Connect` item calls up a dialog box which requests IP and port numbers for the `MDSPlus` server. The `Open` item calls up a dialog box which requests a particular experiment name and shot number. When this information is provided, the directory structure for that shot is downloaded from the server and is displayed as a file tree. When a "file node" of the tree is clicked, the directory path to that file is displayed and a labelled waveform graph is plotted. If there is no data or invalid data then a message is displayed in the graph panel."

Had we not already written this code, the above description could be an informal statement of the requirements of the software from the perspective of a user. It describes how the GUI looks to a user and, in general terms, what it does. But it does not contain any information about how the software needs to be written to interact with the database. It does not address questions such as:

- How is a connection made to the server?
- What is the "language" used to interact with the `MDSplus` server?
- How does the server return a full directory structure?
- How does the server return the plot data?
- What sorts of error-handling should there be in the interaction with the server?

All of these questions are irrelevant to the appearance, and the basic functionality of the GUI and to the appearance of the graphics display. They are a *completely separate concern* to the GUI and graphics display and it makes a lot of sense to put the code that deals with them into an *entirely different subsystem of classes*. Splitting up object-oriented software to enable such a separation of concerns can be achieved using the *facade pattern*. It is one of the most important lessons of this book.

8.1 Facade

Figure 8.1 shows a facade pattern. A whole subsystem of classes has been boxed inside a large rectangle with a little tab on top. (The sUML symbol for a subsystem or package.) A facade class (called `Facade!`) sits on the subsystem boundary. In the figure, a client class outside this subsystem has an association with the *interface* of the facade class (and this interface has also been drawn outside of the subsystem). The idea is that the interface would contain signatures of all of the methods which represent *services provided by the subsystem*. It is the external view, or "face", of the subsystem and the idea of it being a facade in the architectural sense is very pleasing – particularly because design patterns originated from architecture. The facade class implements its interface by delegating much of the processing to classes, `Class1`, `Class2`, ... `ClassN`, inside the subsystem.

The `FacadeInterface` interface is a window onto the subsystem. It describes what services other classes outside of the subsystem can expect. The subsystem itself will also, most likely, need to request services from other subsystems. In most circumstances, these method calls will made by the facade class rather than by other classes deep within the subsystem. Note that, in Java, method calls cannot be made by interfaces because they do not contain any actual code so outgoing calls cannot be made from the interface.

The way that two subsystems communicate using facades is shown in Fig. 8.2. This facade pattern is the fundamental pattern for splitting up your subsystems and keeping them under control. In Sect. 8.3 we will walk through the process of splitting up Java code to properly implement this pattern.

8.2 `EScope0`: A "Do Nothing" Code Refactoring Using Packages

A class diagram for `PreEScope4` is shown in Fig. 8.3. We will now start our restructuring of this code. In doing so we will change the name from `PreEScope` to `EScope`. Our first version of `EScope`, `EScope0`, will start to establish our package structure. The treatment in this section reads like a recipe and readers will get most value from it if they step through the operations being described. It is an easy recipe.

8.2 EScope0: A "Do Nothing" Code Refactoring Using Packages

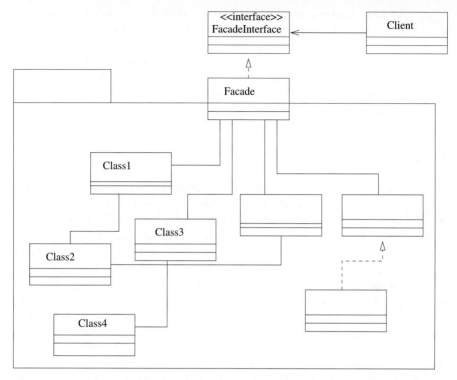

Fig. 8.1. Class Diagram for a hypothetical Facade pattern. The Facade class "wraps up" a subsystem. An external client interacts with classes in the subsystem only through the Facade and its interface.

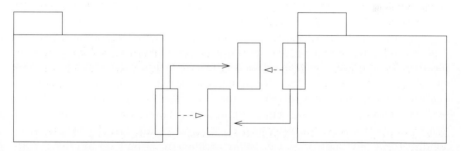

Fig. 8.2. Overview of two subsystems which are associated using the Facade pattern.

116   8 First Facades

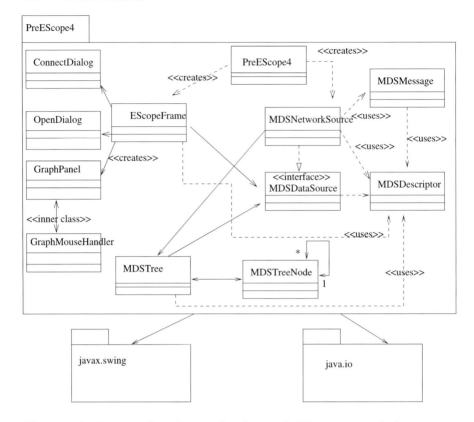

Fig. 8.3. A schematic class diagram for the simple EScope system before restructuring.

Although the `PreEScope4` class diagram might appear to be quite complicated, the linkages between classes already partition the system into two well-defined subsystems. One subsystem is essentially to do with *display* and *control* (through the GUI interface) and the other is to do with *interaction with the data server*. So we will start out with two distinct domains called "gui" and "dataServer" and we will create a directory called `escope0` with subdirectories `guiDomain` and `dataServerDomain`. (Don't do anything with your classes just yet.)

You might think of the mission statements of our two domains as:

guiDomain: Deals with the appearance of all of the GUI panels including the dialogs and the graphics panel.

8.2 EScope0: A "Do Nothing" Code Refactoring Using Packages

dataServerDomain: Handles all communication with the `MDSPlus` server. No `MDSPlus` syntax like "`getnci`" is permitted in any other domain apart from this one.

This is all well and good, but one aspect of the GUI is extremely complicated: the `GraphPanel` is a very large class and, as time goes by, we will want to add even more functionality to it. To make it manageable, we will want to split it up into smaller classes. All of these classes will have a common mission statement to do with *graphics* and it makes much sense to define an additional, `graphics` domain to contain them. So here is a better set of mission statements:

guiDomain: Deals with the overall appearance and coordination of all of the GUI panels. Contains the detailed specification of all GUI components except for the graphics panel.

graphicsDomain: Contains the details of the graphics panel including the mouse event handling.

dataServerDomain: As before.

So, let us create a new subdirectory called `graphicsDomain`. Now we can copy the classes into the appropriate directories but we need to place appropriate `package` and `import` statements at the beginning of each class. For example, every class in the `guiDomain` package will be able to find every other class if it has the header:

```
package guiDomain;
import dataServerDomain.*;
import graphicsDomain.*;
```

The `PreEScope4` classes will be divided up as follows

guiDomain: `EScopeFrame`, `OpenDialog`, `ConnectDialog`
graphicsDomain: `GraphPanel`
dataServerDomain: `MDSDataSource`, `MDSNetworkSource`, `MDSDesriptor`, `MDSMessage`, `MDSTree`, `MDSTreeModel`, `MDSTreeNode`

The main, `EScope0` class, in the top level directory will look something like this:

```
import dataServerDomain.*;
import guiDomain.*;
import graphicsDomain.*;
import javax.swing.*;

public class EScope0
{
    public static void main(String args[])
```

```
    {
        new EScope0();
    }
    public EScope0()
    {
        EScopeFrame gui = new EScopeFrame();
        MDSDataSource dataServer = new MDSNetworkSource();
        gui.initialise(dataServer);
    }
}
```

If you are in this top directory and you type

```
javac EScope0.java
java EScope0
```

then the program should compile and run just fine!

At this stage we have made some progress in managing our software but, to some extent, this progress is an illusion! Every class in each package (domain) is visible to all other classes so we have not really been able to separate concerns: a change in a class in any package might potentially impact on all other classes. We will address this problem in Section 8.3 where we will progressively implement the facade pattern to make the domains more independent. But first a note about a useful utility called "make".

8.2.1 Using Makefiles

A good way of managing the compilation of files on a Unix system, or on Windows using Cygwin, is to use a "makefile". If you copy the following into a file called Makefile in the top, EScope0, directory, then the commands make compile and make run will compile and run your program. make removeall will remove all class files.

```
default: compile run removeall

compile:
        javac EScope.java

run:
        java EScope

removeall:
        rm *.class *~; cd dataServerDomain; rm *.class *~; cd
        ../guiDomain; rm *.class *~; cd ../graphicsDomain;
        rm *.class *~; cd ..
```

Makefiles have a special syntax. They must have tabs after the name of the command, so the line below compile: (which is known as the "command target") starts with a tab. Also be careful not to break lines inside a command target. For example, the target of the `removeall` command must wrap around lines rather than being broken as shown in the listing! Using `make`, you could modify a class in, for example, the GUI domain and then type the following sequence of commands in the top directory:

```
make removeall
make compile
make run
```

We recommend removing all class files before compilation to ensure that your changes have made it through to the latest version of your software. Don't trust `javac` to do this for you! Later as you get more familiar with make, you could organize yourself to only recompile parts of your software.

`Make` is something known as a "build utility". It is widely used to manage large software projects. It is simple if you can remember the funny syntax and put up with the oblique error statements, but be aware that there are other possible build utilities to choose from. Many Java developers recommend another utility called "`ant`".

8.3 `EScope1`: First Implementation of the Facade Pattern

We will now start to refactor `EScope` to incorporate the facade pattern. Before we start, refer back to Fig. 8.3. It is apparent that there are three classes which could serve as facades for our three domains. We shall start our refactoring by renaming these classes

- `EScopeFrame` can be the Facade for the `guiDomain` package. Rename it to `GuiFacade`.
- `GraphPanel` as the Facade for the `graphicsDomain` package. Rename it to `GraphicsFacade`. Move the constructor of this class to the main, `EScope1`, class and pass a handle to it through to the `GuiFacade`. In this way all of the Facades are created by the `main` method of `EScope1`. Then they are initialized by making the necessary connections between domains. Finally, control is passed to the GUI and the main method finishes.
- `MDSNetworkSource` can be the Facade for the `dataServerDomain` package. Change its name to `DataServerFacade`.

Ideally it would be great if only the facade classes needed to import other packages because then we are able to regulate "traffic" into and out of domains more easily. So, now attempt to delete all `import` statements apart from those in the three facades and recompile. Surprisingly, this should work.

8.3.1 Place Facade Interfaces into a Shared Package

According to Gamma et al. [28], the first principle of reusable, object-oriented design is

120   8 First Facades

> *Program to an interface, not to an implementation.*

If interfaces are used to wrap up domains, then the interface to a domain helps to define the requirements of that domain. The interfaces to other domains help to define the services which can be assumed to be delivered by those domains. Implementations of the classes inside the domains, including the facade classes, can be made at will providing the interfaces do not change. In this way, work can proceed on one domain independently of work on the others. Software projects can be split-up and delegated to teams of programmers.

We shall now define interfaces for our facade classes and place them in a separate package, `sharedInterfaces`.

8.3.2 Facade Interface for the GUI Domain

Consider the `GuiFacade` class. It is a very complicated class but it only contains one method which gets called from other domains. This is the `initialise` method which is called from `EScope1`. So the `GuiFacadeInterface` will contain only the following method:

public void initialise (.......)

This way of initializing facade objects will be common to all facades. To make it easier for ourselves, let us adopt the following convention:

> Facade classes will have constructors which contain zero arguments. They will then expose an `initialise` method in their interface which needs to be called in order to properly initialize objects of that class and to associate that domain with other domains.

This style convention is just an attempt to keep our software simple and consistent. A class which wants to use an interface only needs to know the form of the `initialise` method of the interface which that facade implements. No knowledge about the constructor of the actual facade is needed as it will always have no arguments. This rule is aligned with the style conventions we adopted earlier for our demonstration GUIs in Sect. 3.2. Java interfaces do not include the class constructors, so we need to adopt a convention like this one if we wish to be able to advertise how to construct and initialize objects. (Another solution which people adopt is to use a "factory method". This will be described further in Chapter 17.)

8.3.3 Facade Interface for the Data Server Domain

`MDSDataSource` is the interface which is implemented by the `DataServerFacde` class. It is a good candidate for a facade interface for this domain. We should now change its name to `DataServerInterface` and put it into the `sharedInterfaces` directory.

There is one class inside the data server domain which needs to be accessible from outside. This is `MDSDescriptor` which describes the data packet which gets sent to and from `MDSplus`. The GUI domain needs to know about this class in order to request plot data from the server. The class also needs to be directly accessible from several other classes inside the data server domain. At present, it is not immediately obvious what should be done about this class and we shall place it into its own package, `sharedObjects`, which we will allow to be imported by any classes which need it. In the next chapter, in Section 9.4, we shall see how the facade pattern can be "articulated" to deal properly with classes such as this one.

8.3.4 Facade Interface for the Graphics Domain

`GraphicsFacade` is a very large class but it only needs to expose two methods to other domains: `initialise`, called by the main method, and `setGraphData`, called by `GraphicsFacade` when a user clicks on a dataset in the file tree. But the construction of an interface for this class is complicated by a subtle problem. `GraphicsFacade` also extends the Swing `JPanel` class and it is passed to other components in the GUI which assume that it is a `JPanel`. We need to be able to advertise that this class will be a `JPanel` but because `JPanel` does not implement a `JPanelInterface` we cannot just let our `GraphicsFacade` implement two interfaces. Instead, we need to have an abstract facade class which both implements the `GraphicsFacadeInterface` and extends `JPanel`:

```
package sharedInterfaces;
import javax.swing.*;
public abstract class AbstractGraphicsFacade extends JPanel
                 implements GraphicsFacadeInterface {}
```

8.3.5 Our Final Product

If you have been refactoring `PreEScope4` following this recipe then you might find that a bit of "fiddling" will need to take place before all of the links have been made successfully and you have the correct structure in place. You should check that there are no unnecessary `import` statements remaining in the software. For example, ideally the three facade classes should only need to import the `facadeInterfaces` package. In fact, as discussed in Section 8.3.3 above, we have a problem with the shared `MDSDescriptor` class so we can expect that several classes might need to import its special `sharedObjects` domain.

The final class diagram for the reconstructed system is shown in overview in Fig. 8.4 and for the three domains in Figs. 8.5 and 8.6. The three, "facaded" domains are quite neatly structured and loose coupling is provided by the `sharedInterfaces` package. As remarked above, there is some residual messiness in the way that the `MDSDescriptor` class has been handled.

In the present design, it has been placed in a sharedObjects package and made accessible to classes which need it. We have drawn a "uses" package association from the data server domain to this package to show that classes inside this domain may have linkages to it. This messiness will be addressed in Section 9.4.

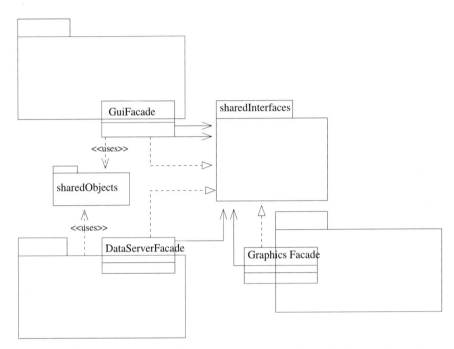

Fig. 8.4. A schematic class diagram for EScope1. Dependencies on Java Swing and IO packages are not shown. Shared interfaces and abstract classes are gathered into the sharedInterfaces package. The MDSDescriptor class is located in the sharedObjects package.

After all of the above changes, the main EScope1 class looks like:

```
import sharedInterfaces.*;
import dataServerDomain.*;
import guiDomain.*;
import graphicsDomain.*;
import javax.swing.*;

public class EScope1
{
    public static void main(String args[])
    {
        new EScope1();
```

9
Adapter

Imagine that you are constructing a software system and are assuming that one particular class in this system implements a certain interface. Then someone supplies you with a concrete class (perhaps as a Java class file) which does exactly what you want it to do except that it does not implement the interface that you have been designing for. In this situation, the "adapter" pattern can often be used to integrate the supplied class with your software. Its idea is to construct a special "adapter class" which implements the interface that your code expects (the target interface). The adapter class is usually one of two types: it might make calls to an object of the supplied class (the "object adapter" pattern) or it can use multiple inheritance to extend the supplied class as well as implementing the interface that your code requires (the "class adapter" pattern).

9.1 Object Adapter Pattern

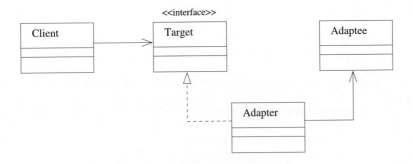

Fig. 9.1. Object Adapter pattern.

A class diagram for the object adapter pattern is shown in Fig. 9.1. The object adapter class implements an interface expected by the client. It maintains an association with the supplied, adaptee, class and its implementation of some methods in the target interface will delegate to methods in the adaptee class.

A very simple example of an object adapter is one in which the names, and signatures, of methods in the adaptee class become adapted to those expected by the required interface. If, for example, a fictitious "`BottleModel`" interface required a method `pourOut(int standardDrinks)` and the `SuppliedBottle` class contained a method `pour(float volume)` then an adapter class, "`BottleAdapter`", would implement `BottleModel` and its constructor would need to create an object, "`myBottle`" of type `SuppliedBottle`. Its `pourOut` method might delegate to this object as follows:

```
public void pourOut( int standardDrinks )
{
    float volume = standardDrinks * volume_per_drink;
    myBottle.pour( volume );
}
```

A more sophisticated object adapter will be used in our next revision of `EScope`. It will go something like this: Imagine that an object containing read-only data gets passed to a subsystem of classes. For example, the object might contain user-specified options for the appearance of graphs. It might be written by a subsystem of classes dealing with user interaction and get passed to a subsystem of classes dealing with graphics. Inside the graphics subsystem, it might be convenient to add some more graphics parameters to this object. This could be done by associating the original object with another which represents the enhanced set of graphics options and which delegates to the original object using the object adapter pattern. The `GraphOptions` class which defines this enhanced object could look like the following code excerpt:

```
public class GraphOptionsAdapted implements
                          GraphOptionsAdaptedInterface
{
    private GraphOptionsSuppliedInterface suppliedObject;
    private int otherParameter1, otherParameter2;
    private double otherParameter3, otherParameter4;

    public GraphOptions( GraphOptionsSuppliedInterface opt )
    {
        suppliedObject = opt;
        otherParameter1 = default1;
        otherParameter2 = default2;
        ...
    }
    // accessor methods for user supplied options
    public int getUserParameter1()
```

```
                    { return suppliedObject.getParameter1();}
    public int getUserParameter2()
                    { return suppliedObject.getParameter2();}
....
// accessor and mutator methods for other options
pubic void setOtherParameter1(int param)
                    { otherParameter1 = param;}
    public int getOtherParameter1()
                    { return otherParameter1;}
...
}
```

9.2 Class Adapter Pattern

A class diagram for the class adapter pattern is shown in Fig. 9.2. In the class adapter pattern, the adpter class will, typically, implement the required interface whilst extending the supplied, adaptee, class. This is very convenient for some applications. For example, the "pourOut" example of the previous section would be simpler if the adapter class had extended the SuppliedBottle class so that the pour method could be called directly:

```
public void pourOut(int standardDrinks)
{
    float volume = standardDrinks*volume_per_drink;
    pour(volume);
}
```

The class adapter can have a even simpler implementation in programming languages which allow true multiple inheritance. In this case the adapting class would simply extend two superclasses.

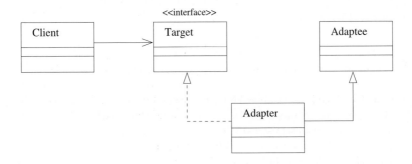

Fig. 9.2. Class Adapter pattern.

9.3 Are Object Adapters Better Than Class Adapters?

There are several advantages in using the object adapter pattern instead of the class adapter. One is that an object adapter can adapt a whole hierarchy of adaptee classes rather than being tied to a particular class: The instance field for the adaptee class could be an interface of a super-class. The constructor could then be passed an object of any concrete class implementing this interface.

Another advantage of the object adapter is that of run-time operation versus compile-time inheritance. An object adapter is often constructed by passing an adaptee object at run-time, and that object can be used quite simply. This is not the case with the class adapter pattern.

In general, the decision to use an object adapter rather than a class adapter can be viewed as a choice between *object composition* and *inheritance*. In Section 8.3.1 we discussed the first principle of object-oriented design. According to Gamma et al. [28], The second principle of object-oriented design is

Favor object composition over inheritance.

The reasons for this are many and varied. Some software engineers are convinced that an over-use of inheritance can lead to software becoming difficult to maintain and reuse. Favoring object composition makes your classes and class hierarchies small and well focussed. Although object composition is meant to be a good thing, it is more work to code an object adapter pattern rather than a class adapter because all method calls must be explicitly delegated.

The object adapter pattern will be of practical use for us in the next refactoring step for EScope and we will meet it again several times in the rest of this book.

9.4 EScope2: Sharing Graph Data and Graph Options Between Domains

We will now proceed to improve the structure and functionality of EScope using the object adapter pattern. Our starting point is the observation of the messiness in EScope1 caused by the placement of the MDSDescriptor class in a package which is imported by several classes inside the data server domain (Section 8.3.5). We will use interfaces and the adapter pattern to police access to this class more effectively.

MDSDescriptor objects are parcels of data. It is wise to parcel related data together in objects. Indeed, this is one of the big points of object oriented programming! But EScope1 is dealing with graph data in a messy fashion. Consider what happens when a user clicks on a file containing data. Ignoring error handling, the steps in the procedure are:

9.4 EScope2: Sharing Graph Data and Graph Options Between Domains

1. The `GuiFacade` object contacts the data server 4 times to download `MDSDescriptor` objects.
2. The X and Y data arrays and X and Y labels are unpacked from the `MDSDescriptor` objects.
3. The `GraphicsFacade` object is sent the data arrays and other information needed to construct a plot with the method call:

    GuiFacade.this.graph.setGraphData(xVals, yVals,
          xUnits, yUnits, shortFileAddress,
                isGraphData, isDataError);

This procedure is messy because the `GuiFacade` needs to know about the structure of the `MDSDescriptor` objects in order to unpack the data. It would be much more satisfactory if the data could by-pass the `GuiFacade` and be sent directly from the data server to the `GraphicsFacade`. This is one improvement we will seek to make in `EScope2`. (It will have the additional, useful effect of moving the `MDSDescriptor` class into the data server domain which is convenient because it is used by several classes there.) But we will still need to come up with a mechanism to let the graphics facade know about the structure of the object which contains the data to be plotted. If the interface for this object were to be published in a shared domain, called **sharedDataInterfaces**, then this would meet our requirements. We shall do this, but will be careful to only publish *accessor* methods for the data in this interface. The data will be written by the data server facade and read by the graphics facade. Here is an example of the way the interface could look. As well as simple accessor methods, additional methods have been added to compute useful quantities such as the minimum and maximum X and Y data values:

```
package sharedDataInterfaces;

public interface GraphDataInterface
{
    // utility methods
    /** Return length of y data array*/
    public int getLength();
    /** Returns the smallest x axis data point*/
    public double getMinX();
    /** Returns the largest x axis data point*/
    public double getMaxX();
    /** Returns the smallest y axis data point*/
    public double getMinY();
    /** Returns the largest y axis data point*/
    public double getMaxY();

    // other get methods
    public double[] getX(); // X data
    public double[] getY(); // Y data
    public String getXLabel();
```

```
    public String getYLabel();
    public String getTitle();
    public boolean getIsGraphData();
    public boolean getIsDataError();
}
```

9.4.1 Passing Graph Options from the User Interface

Consider an enhancement of EScope which will allow users to specify some parameters determining the appearance of a plot. The interface for the object which will store these parameters would lie in the GUI domain. Some additional GUI components will be needed and these might consist of an extra menu item together with a specialized dialog box. It is wise to imagine that this dialog might become quite ambitious over time because users might demand that they could make more and more cosmetic modifications to graphics. It is also evident that not all of the parameters needed to describe a plot would be under the control of users as some parameters would be system defaults and would be set by the graphics facade.

In a similar way to the interface for the plot data, we can advertise an interface of accessor methods corresponding to the plotting parameters under control of the user. The object conforming to this interface would be written by the graphics facade and be passed to the graphics facade where it would get wrapped up as a new "GraphOptionsInGraphics" object using the object adapter pattern. The corresponding class diagram looks something like Fig. 9.3.

9.4.2 An *Articulated Facade*

The example of constructing a data object and passing it to another domain where it gets wrapped up and used in that domain seems to be quite general and it gets around the problem of multiple linkages from classes to an object outside of their particular domain. It can be thought of as representing a "data channel" of communication between the two domains. We have decided to give this refinement of the facade pattern a special name: the "articulated facade". We think that this is a reasonable name because it implies that the simple facade pattern has been *articulated*, in some way, to become clearer or more logical.[1] It is also a neat name because articulated facades are associated with architecture which is the field which first employed the notion of design patterns!

A real, architectural, articulated facade is shown in Fig. 9.4.

---

[1] Our articulated facade pattern is not an official term. In the culture of design patterns, a new pattern definition should be subjected to a process of review and shepherding. Pattern Languages of Programming (OLOP) conferences have been held around the world to facilitate this process.

9.4 EScope2: Sharing Graph Data and Graph Options Between Domains

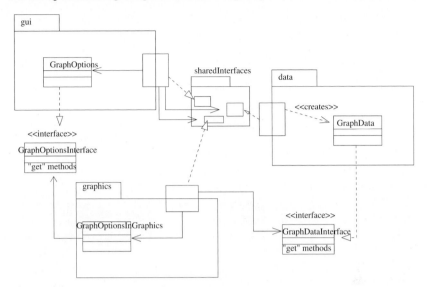

Fig. 9.3. A `GraphOptions` object can be passed from the GUI domain to the graphics domain and then wrapped up by a `GraphOptionsInGraphics` object using the object adapter pattern. A `GraphData` object can be passed from the data server domain to the graphics domain by advertising its interface. This object will be written in the data server domain and read in the graphics domain but an adapter pattern is not needed (yet!) because the object is not wrapped up inside the graphics domain. Both data channel interfaces can be grouped together inside a `sharedDataInterfaces` package. Note that the domain facades and their interfaces are unlabeled in this figure.

9.4.3 Our Final Product

The refactoring work needed to produce `EScope2` follows the discussion of the previous sections and starts by defining interfaces for the graph data and graph options objects. These can be moved to their own "`sharedDataInterfaces`" package. The `MDSDescriptor` class gets moved into the data server package and the overall structure is shown in Fig. 9.3. The object adapter pattern contributes to our articulated facade structure in the wrapping of the graph options object when it enters the graphics domain. Additional functionality has been added to allow users to specify colors for the graph labels, graph background and the plot itself. Further details are discussed below and the software is available on the CD included with this book.

9.4.4 Data Server Domain

As noted above, the data server domain is altered by moving the `MDSDescriptor` class into it and doing away with the separate `sharedObjects` package. An additional method:

Fig. 9.4. A real, architectural, articulated facade!

```
public GraphDataInterface getPlotData
               (String shortFileAddress)
```

is added to the `DataServerFacade` class. This method contacts the server repeatedly to assemble an object of graph data. Its listing is given in Appendix E.4.1 on page 300. It is called from `GuiFacade` in order to pass the graph data object to the graphics domain:

```
GuiFacade.this.graph.setGraphData
          (database.getPlotData(shortFileAddress));
```

The `GraphDataInterface` interface is placed in a new `sharedData-Interfaces` package.

A class diagram for this domain is shown in Fig. 9.5.

9.4.5 GUI Domain

The GUI domain needs to contain extra classes to solicit user specification of graphics options. We have chosen to start with options to specify different colors for various parts of the graph. This gives rise to a complication: colors can be stored as `java.awt.Color` objects but they need to be mapped to an array of strings so that their names can be read from a dialog. They also need to be mapped to integers if they are to be selected from a Swing component such as a `JComboBox`. These mappings can either be hidden inside the GUI domain, which would make the shared interface compact and elegant, or they

9.4 EScope2: Sharing Graph Data and Graph Options Between Domains

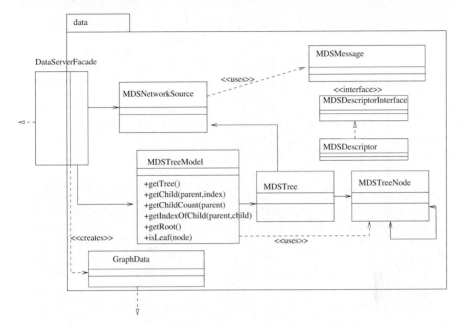

Fig. 9.5. Class diagram for the data server domain for EScope2.

can be advertised in the shared interface. We choose to advertise them in the interface so that they can be easily read and modified:

```
package sharedDataInterfaces;
...
/** Stores the options settings for the graph panel */
public interface GraphOptionsInterface
{
    public static final int numColors = 13;
    public static final int BLACK = 0;
    public static final int WHITE = 1;
    public static final int RED = 2;
    public static final int BLUE = 3;
    public static final int ORANGE = 4;
    public static final int GRAY = 5;
    public static final int LGRAY = 6;
    public static final int DGRAY = 7;
    public static final int MAGENTA = 8;
    public static final int GREEN = 9;
    public static final int YELLOW = 10;
    public static final int CYAN = 11;
    public static final int PINK = 12;
    public static final    String[] colorNames =
       {"Black","White","Red","Blue","Orange","Gray",
```

```
              "L. Gray","D. Gray","Magenta","Green","Yellow",
                              "Cyan","Pink" };

    /** returns a Color object corresponding to index i */
    public Color selectColor(int i);
    /** returns a string colour name
                            corresponding to index i */
    public String selectColorName(int i);
    public int getTitleColor();
    public int getGraphColor();
    public int getLineColor();
}
```

If you play with this version of the software on the accompanying CD, you will note that we have also chosen to make the new options-dialog *modeless* so that it does not block the main GUI. This is in order to implement an "Apply" button which lets new options affect a graph's appearance without the dialog box disappearing. Corresponding calls to a new method, `setGraphOptions`, of the graphics facade need to be made and this new method needs to be advertised through the shared graphics interface.

A class diagram for this domain is shown in Fig. 9.6.

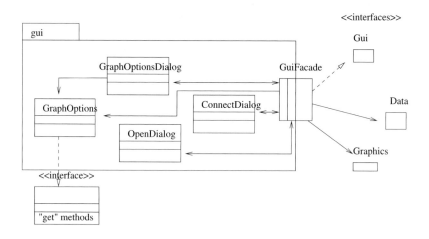

Fig. 9.6. Class diagram for the GUI domain for `EScope2`.

9.4.6 Graphics Domain

For the `GraphOptionsInGraphics` class, we chose to augment the user specified graphics options with default values for: the number of significant figures; some font parameters; and some plotting strokes. This can be seen in the first part of this class in the following listing:

9.4 EScope2: Sharing Graph Data and Graph Options Between Domains

```java
public class GraphOptionsInGraphics
{
    private GraphOptionsInterface graphOptions;
    private int sigFigs, titleFontHeight,
            axisFontHeight, tickFontHeight, tickPix;
    private double textFac;
    private Font titleFont, axisFont, tickFont;
    private double maxNormalRange=10;
    private double minNormalRange=1;
                        // (Do not use sci notation inside
                        //   of normal range for ticks)
    private BasicStroke lineStroke, axisStroke, titleStroke;

    public GraphOptionsInGraphics(
                            GraphOptionsInterface options)
    {
        graphOptions = options;
        sigFigs = 5;
        titleFontHeight = 16;
        axisFontHeight = 12;
        tickFontHeight = 10;
        textFac = 0.8; // (ascent+leading)/(text-height)
        tickPix = 4; // number of pixels for tick length

        titleFont = new Font(null, Font.PLAIN,
                                        titleFontHeight);
        axisFont = new Font(null, Font.PLAIN, axisFontHeight);
        tickFont = new Font(null, Font.PLAIN, tickFontHeight);

        // stroke for waveform
        float strokeWidth = 1.0f;
        float[] solid = {12.0f,0.0f};    // Solid line style
        lineStroke = new BasicStroke( strokeWidth,
                            BasicStroke.CAP_SQUARE,
                            BasicStroke.JOIN_MITER, 1.0f,
                            solid, 0.0f );
        // stroke for axes and ticks
        strokeWidth = 0.25f;
        axisStroke = new BasicStroke( strokeWidth,
                            BasicStroke.CAP_SQUARE,
                            BasicStroke.JOIN_MITER, 1.0f,
                            solid, 0.0f );
        titleStroke = axisStroke;
    }
}
```

The graphics facade is still too monolithic and we will address its refactoring in the next two chapters. As we split it up using design patterns, we will find that `GraphOptionsInGraphics` objects can be easily associated with specialized classes which plot parts of the graph. It will also turn out to be convenient to augment the graph data objects to store additional information to do with the mapping between real space and pixel coordinates and we will use the object adapter pattern to do this.

10
The Template Pattern

Object-oriented programming texts sometimes stress the primacy of data over processing. Classes are meant to be designed to encapsulate data and to locate methods which act on that data close to the data itself. In this view, the idea of having classes which just encapsulate methods would not seem to be a good thing. But in practice it is often desirable to encapsulate methods without any corresponding data. Otherwise our classes can just become too big and too rigid. One familiar example of a class which encapsulates methods is the `java.lang.Math` class which is a library of useful mathematical methods.

There are several design patterns from Gamma et al.'s "behavioural" [28] grouping which explain how methods can be encapsulated in subsystems of classes. In this chapter we will discuss one of these: the template pattern. We will use this pattern, together with some additional refactoring, to split up the `GraphicsFacade` class into a number of smaller classes whose purpose is to encapsulate algorithms rather than data.

10.1 Pattern Description

The idea of the template pattern is that a complicated algorithm can sometimes be broken up into several parts. One part of the algorithm is common to all of its implementations. In the template pattern, this part of the algorithm is taken out and put into a method of a parent superclass and it is often tagged as being *final*. The other, implementation specific, parts of the algorithm are located in methods which are often declared as *abstract* in the superclass. These abstract methods can then be implemented in different ways in the subclasses.

By using the template pattern, you do not need to keep re-coding the entire algorithm each time you define a new class. In refactoring our case study to produce `EScope3` we will use this pattern to encapsulate the `drawXAxis` and `drawYAxis` methods of the `GraphicsFacade` class.

An issue with the template pattern is that it is sometimes necessary to provide a default for the implementation-specific parts of an algorithm. If these methods are abstract in the superclass then the default implementation must be located in a particular concrete subclass and this must be made clear to client classes. Alternatively the superclass methods can all be concrete and the superclass then needs to document which of them need to be overriden by subclasses.

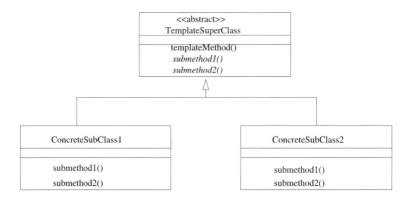

Fig. 10.1. In the template pattern, implementation-specific parts of an algorithm are declared as abstract in the superclass and are implemented in the concrete subclasses.

10.2 EScope3: Splitting up the Graphics Facade

There is quite some work involved in this refactoring step and it can profitably be set as a student assignment (only part of the exercise involves the template pattern, but this is a quite challenging part). The end-point is to have separate classes which deal with drawing the various parts of the graph: the axes, the labels, the waveform and any messages. Along the way, it makes sense to extend our implementation of the articulated facade pattern by constructing a `GraphDataInGraphics` class which can be used to store information relating to the graph data which is additional to what is passed through from the data server domain. This class and the `GraphOptionsInGraphics` class will be associated with a new class, `GraphMetrics`, which will encapsulate the logic to transform from data coordinates to graph pixels. Because `GraphMetrics` will now provide the access to both the graph options and graph data objects from the graphics facade, it will also be tasked with other, "book-keeping", responsibilities such as accessing and modifying data in these objects.

10.2 EScope3: Splitting up the Graphics Facade

If we refer to the discussion of Sections 3.3 and 4.2, and the listing of `Plotter2` in Appendix D.3, then we can recall that the plotting `paintComponent` method contains the following steps:

- find the size of the graphics window,
- paint the background,
- set the pen stroke,
- find the size of the border around the graph region,
- if there is no data or bad data then write a message and stop,
- perform scaling from data coordinates to pixels,
- set up arrays of scaled, filtered data points, guard against divide by zeros in the data arrays,
- plot the curve,
- draw the mouse cursor,
- draw the X axis,
- draw the Y axis, and, finally,
- draw the axis labels.

Compare the listing of Section D.3 with the following excerpt from the `GraphPanel` class of `EScope3` which achieves all of the above and more:

```
public void paintComponent(Graphics g)
{
    // Cast the graphics object to Graph2D
    Graphics2D g2 = (Graphics2D) g;
    // Get plot size
    Dimension size = getSize();
    int xSize = size.width;
    int ySize = size.height;
    // find border size
    int borderSize;
    if (metrics.getGraphOptions().getDefaultBorder())
    {
        borderSize = calcBorderSize(g2);
        metrics.getGraphOptions().setTickPix(4);
    }
    else
    {
        borderSize=0;
        metrics.getGraphOptions().setTickPix(0);
    }
    // Set graph and border size in metrics
    metrics.setXSize(xSize);
    metrics.setYSize(ySize);
    metrics.setBorderSize(borderSize);
    // Set background Color
    g2.setColor(metrics.getGraphOptions().
        selectColor(metrics.getGraphOptions().
                                    getGraphColor()));
```

```java
            g2.fill(new Rectangle2D.Double(0,0,xSize,ySize));
            // If no data or bad data then
            //write messages and break
            if (!metrics.getIsGraphData())
            {
                DrawCentredMessage drawMess =
                                    new DrawCentredMessage();
                String message = " No data";
                if (metrics.getIsDataError())
                    message = "Unexpected data. "
                                +" See console message.";
                metrics.setMessage(message);
                drawMess.draw(metrics,g2);
                return;
            }
            drawWaveform.draw(metrics,g2);
            drawXAxis.draw(metrics,g2);
            drawYAxis.draw(metrics,g2);
            drawLabels.draw(metrics,g2);
            drawCursor.draw(metrics,g2);
}
```

This version of the algorithm is much more compact and elegant. It delegates the drawing parts of the graph to draw methods from the following classes:

- `DrawWaveform` to draw the waveform line itself,
- `DrawXAxis` to draw the X axis,
- `DrawYAxis` to draw the Y axis,
- `DrawLabels` to draw the axis labels,
- `DrawCursor` to draw the mouse cursor, and
- `DrawCentredMessage` to draw an error message if the data is not present or contains unexpected data.

For reasons which will be explained in the next chapter, we have chosen to have uniform signatures for each of these draw methods. Two arguments are passed to them: the graphics context and an object of type `GraphMetrics`. This latter object will enable all of the graph data and graph options to be accessed and the relevant transformations to be made. It has associations with both the `GraphOptionsInGraphics` and the `GraphDataInGraphics` objects.

In `EScope3` we have also implemented two additional options for the display of the graph. The first is to turn the border region on or off. The second option is to choose the algorithm to locate the axis tick-marks and tick labels to be either the nice-numbers algorithm discussed in Section 4.4 or a more straight-forward division into equal bins.

Important parts of this version of `EScope` will remain unchanged for the next two refactoring steps and the following discussion can be illustrated with respect to the listing of `EScope4` in Appendix E on page 295.

10.2.1 The `GraphData` and `GraphMetrics` Classes

Referring again to the `paintComponent` listing for `Plotter2` in Appendix D.3, it is apparent that the efficiency and elegance of the code could be enhanced by moving the part of the algorithm which computes the maximum and minimum values of the graph data to a separate class. The efficiency speed-up comes from the fact that most graph repainting will occur because the window has been resized or because the graph plotting options have been changed and these are conditions which do not necessitate that the data arrays be searched to find maximum and minimum data values for transforming to pixel coordinates. So these values can be computed once and stored somewhere. It is reasonable to expect that they should be associated with the `GraphData` object and, hence, a further extension of our *articulated facade* pattern of the previous chapter is in order: The new class will be called `GraphDataInGraphics` and it will have an association with the `GraphData` object passed from the data server domain. A listing for the `GraphDataInGraphics` class can be seen in Appendix E.5.2 on page 320. (Note that, in this listing, we also have also added a guard against dividing by zero for the X-data.)

Computations of the transformation between data coordinates and pixel coordinates will use information from the graph options object as well as the graph data object. These computations will need to be made for the complete, and possibly filtered, data arrays (for drawing the waveform) as well as for individual pixels (for drawing the mouse cursor). So we would expect a `GraphDataInGraphics` class to contain methods such as `getPixelArray`, to return an array of X and Y pixel values for a waveform, as well as methods such as `getXPixel` or `getXValue` to transform individual data points. This can be carried out by a new "GraphMetrics" class. (The corresponding methods are those in the `GraphMediator` class of `EScope4` as shown in Section E.5.3 on page 324.)

10.2.2 Drawing Individual Graph Components

As remarked above, the draw methods for the individual graph components accept a `GraphMetrics` object (called "`metrics`" in the listing) as well as the graphics context. These graph components are drawn onto the graphics context in pixel coordinates, so their drawing methods start out by retrieving the graphics window size, and the maximum and minimum data ranges, in order to be able to perform the transformation to pixels. (Because the `GraphMetrics` object is passed in a method, rather than having a permanent association, these graph component classes "use" `GraphMetrics` and the linkages are shown as dashed lines in Fig. 10.2.)

Each of the individual drawing classes encapsulates a part of the plotting algorithm. They are not associated with any data apart from the objects passed into their draw methods. But their use is not as straight-forward as it might seem! It is important that both of the X and Y axis-drawing methods be executed before the `DrawLabels` draw method is called. This is because part of the calculation of the axis labels is to determine a power of 10 to use in the scientific notation. This power of 10 needs to be included in the label for each axis; it gets stored in the `GraphDataInGraphics` object and is retrieved by the `DrawLabels` draw method. Because of this order dependence, the implementation of these drawing classes is not completely satisfactory and we will return to them in the next chapter.

10.2.3 The Template Pattern for `DrawAxesTicks`

Before this revision, the axis-drawing methods have been quite complicated. A large amount of code is shared between the `drawXAxis` and `drawYAxis` methods and they run the risk of becoming incompatible with each other if changes are made to one of them. Separating this common code and placing it into an abstract class is an example of the template pattern. We add the `niceNumbers` method to this superclass because it is used by both axis-drawing classes and no others. We also add code to use a more simple, straightforward binning of the axes if that option is specified. We call this a "flexible numbers" option.

The template pattern is greatly facilitated by the use of *protected* variables. These are variables which are only visible to classes which are part of the one inheritance hierarchy. Allowing the implementing subclasses access to protected variables in the template superclass allows us to reduce the size of the parameter lists in method calls. Some authors caution against the use of the `protected` declaration in Java because protected variables also have *package* visibility, i.e. they can be accessed by other classes located in the same package (by default the same computer directory). We try to restrict their use.

Our refactored, axis-drawing classes turn out to be similar to those shown in the `EScope4` listings in Appendix E.5.1 on page 306.

10.2.4 Our Final Product

The refactored graphics domain is shown in Fig. 10.2. This is no longer a monolithic class. But this class diagram is messier than we would like and it should be considered as being "work in progress". We will tidy it up in the next chapter where we will gather, or chain, our drawing classes together using the decorator pattern.

Because we have extended our articulated facade structure to include the graph data object, the overall class diagram for the `EScope3` software system differs from Fig. 9.3 in a fairly obvious way. Instead of having a direct link from the `GraphicsFacade` to the `GraphData` class, instances of this class are now passed into the graphics domain using the object adapter pattern, with the adapter class being `GraphDataInGraphics`.

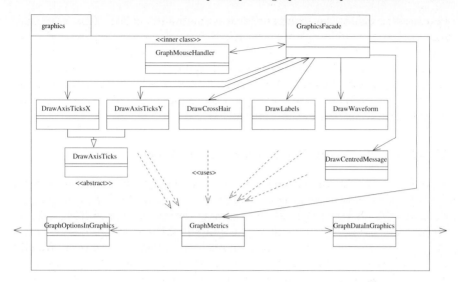

Fig. 10.2. Class diagram for the graphics domain for EScope3.

11
Decorator

We have already met the decorator pattern when we described the Java Input-Output framework in Section 2.2. It is characterized by a chaining together of certain classes to produce a more complicated class which exhibits a certain behavior. In the IO framework, these classes can be chained together to

- associate data streams with a number of different external file types,
- associate streams with Java programs, and
- process the data in different ways as it is streamed.

Chaining together IO classes enables a large number of composite behaviors to be constructed from a smaller number of fundamental classes.

It is hard to resist using the decorator pattern for `EScope` because the idea of "decorating" a graph resonates with that of adding annotations such as axes, ticks and labels. In fact, it does turn out to be quite a elegant pattern in our context. It also manages the various "draw" classes which were introduced in the previous chapter and it enables them to be easily extended.

11.1 Pattern Description

One possible representation of the decorator pattern is shown in Fig. 11.1. The abstract decorator superclass maintains a reference to a component which is of the same type as itself. Requests that the decorator receive can be forwarded to the component and to that component's component, and so on down the chain.

The following code excerpt, from the `GraphDecorator` abstract class (in `EScope4`), shows how the pattern might be implemented. Objects of concrete classes which extend this class need only to implement the `drawThis` method. The `draw` method will check whether there is a link to a component decorator object and, if so, will call the component's `draw` method before calling its own `drawThis` method. Note that there is a link to a new class called `GraphMediator`. This will be discussed below in Section 11.2.2.

148   11 Decorator

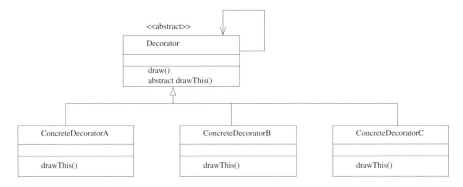

Fig. 11.1. Decorator pattern.

```
package graphicsDomain;
import java.awt.*;

//Implementation of the decorator pattern
public abstract class GraphDecorator
{
    private GraphDecorator decorator=null;
                          //decoration chain component
    protected GraphMediator med;

    public GraphDecorator(GraphMediator m) { this.med =m;}
    public void setDecorator(GraphDecorator decorator)
    {
        this.decorator = decorator;
    }
    //Draw component first, then draw this decoration
    public void draw(Graphics2D g2d)
    {

        if (decorator != null) decorator.draw(g2d);
        drawThis(g2d);
    }
    abstract public void drawThis(Graphics2D g2d);
}
```

Although we do not do this in our example, decorators are usually chained together at the time they are constructed and a nested pattern of constructors is a hint that a decorator pattern may be at work. For example, we constructed an object to read unformatted data from a file using `DataInputStream` in Section 2.2:

11.2 EScope4: Adding Zoom and Grab Options Using the Decorator Pattern

```
DataInputStream inData =
   new DataInputStream(
   new BufferedInputStream(new FileInputStream("data")));
```

This chaining of the classes `DataInputStream`, `BufferedInputStream` and `FileInputStream` produced an object which established a stream between the program and a particular file. This stream could then be used to read buffered data from the file. An alternative to using the decorator pattern in this case would have been for the Java API to define particular subclasses of `FileInputStream` which had the required functionality. But this would have resulted in a more rigid overall IO framework as well as a large number of classes with a complicated inheritance hierarchy.

A decorator object can also be applied dynamically. It can be removed and could also be added more than once. We will see examples of this in our development of `EScope` in this chapter.

11.2 EScope4: Adding Zoom and Grab Options Using the Decorator Pattern

This refactoring step marks a milestone in our development of `EScope`. With it, much of our basic architecture is complete. The decorator pattern fixes up a problem which was left dangling at the end of the previous chapter. It also enables us to elegantly extend the software to change the way that a graph is drawn. Because it is an important reference point for the discussions in this text, much of the code is reproduced in its entirety in Appendix E on page 295.

In Chapter 10 we showed a `paintComponent` method which drew an annotated graph using calls to a number of `draw` methods:

```
drawWaveform.draw(metrics, g2);
drawXAxis.draw(metrics, g2);
drawYAxis.draw(metrics, g2);
drawLabels.draw(metrics, g2);
drawCursor.draw(metrics, g2);
```

We mentioned, in Section 10.2.2, that there was a constraint on the ordering of these method calls. It would be convenient to hide this constraint inside a method which constructed a decorator chain which could be drawn by calling one draw method. This would also give us the flexibility of constructing different composite decorator chains for different graphs.

The reason for our uniform method signatures for the draw methods should now be apparent: they make it relatively straight-forward to convert these classes from their `EScope3` versions into a decorator pattern. Their listings can be found in Section E.5.1 on page 306.

11.2.1 Grab and Zoom

In refactoring to `EScope4` we take the opportunity to demonstrate the elegance of the decorator pattern by adding two new options for manipulating the graph using the mouse. The first is a "grab and drag" option. When this option is specified, clicking and dragging a graph will move the position of the waveform and the axis ticks and labels will adjust themselves accordingly. The second option is a zoom: clicking and dragging the mouse will draw a box about a section of the graph waveform. Releasing the mouse button will blow up that section of the graph to occupy the entire graph. For both options, double clicking the mouse will reset the graph to its original display.

The mouse behaviors needed for these two new manipulation options are implemented in the `GraphMouseHandler` inner class of `GraphicsFacade`. A new decorator class also needs to be constructed to draw the zoom box. A key observation is that the transformations between data and pixel coordinates involve parameters which measure the maximum and minimum extents of data in the x and y directions. The default is for these to be read from the data arrays, but it is also possible to set them individually. The new graph options will work by calculating new values of these maximum and minimum extents and then recalculating the transformations. The `GraphMouseHandler` listing can be found in Section E.5.4, on page 330, and the zoom decorator is in Section E.5.1 on page 318. There is no decorator for the grab option.

11.2.2 A Mediator Emerges

The class diagram for the graphics domain of `EScope3`, in Fig. 10.2, turned out to be messier than we would have liked due to the placement of the "draw" classes. In that version the draw classes were constructed and called from the graphics facade but they also made use of a graph metrics object in order to perform the coordinate transformations and to access fields from the `GraphOptions` and `GraphData` objects. This can be replaced by a more elegant structure by pushing the decorator chain down to be "below" the graph metrics class. If the graph metrics class is responsible for constructing and managing the decorators, then the main `paintComponent` method of the graphics facade becomes even simpler than it was in `EScope3`. In fact, there is an increased need to move the decorators away from the facade because of the additional coding needed to instantiate different decorator chains and because of the additions to the `GraphMouseHandler` inner class of the facade. These new additions run the risk of making the graphics facade too big and unmanageable.

As the graph metrics class becomes not only responsible for the coordinate transformations but also for managing the decorator classes, it starts to look more and more like a mediator class. Even though, as discussed in Section 8.4, the mediator is meant to be a coordinating class rather than one which contains processing on its own, we can tolerate the retention of the coordinate transformation methods in our, "`GraphMediator`" class (renamed

11.2 EScope4: Adding Zoom and Grab Options Using the Decorator Pattern

from `GraphMetrics`) because they emerge from an interaction between the graph options and graph data objects and because the size of this class is yet to become too unmanageable. (If, in the future, this class did grow to be "too big" then we could consider moving these methods to a new, more narrowly defined, metrics class and keep the graph mediator class to manage the class interactions.)

The `GraphMediator` class is shown in Section E.5.3 on page 324. It contains associations with objects from all of the concrete decorator classes and it constructs and returns composite decorators using three methods:

- `createDecorator`
- `concatenateDecorators`
- `getDecorator`

The `createDecorator` method is called whenever a new graph manipulation option is chosen, it selects the correct combination of decorator objects to chain together and chains them together in the appropriate order. The `concatenateDecorator` method performs the chaining and the `getDecorator` method returns a link to the composite decorator object. (Note that `concatenateDecorator` accepts a variable number of arguments which is new to Java from Java 1.5.) Listings of all three methods follow. The `createDecorator` method is a type of *factory method* which is a pattern which will be discussed in Chapter 17.

```java
public void createDecorator()
{
    /*Build the decorator chain.
      The elements in the chain define what
      is going to be displayed.

      If no data or bad data then the decorator chain
      is only composed of a DrawCentredMessage object
      containing an error message.*/
    if (graphData == null || !isGraphData)
    {
        if (graphData != null && isDataError)
            warningDecorator.setMessage(
                "Unexpected data. See console message.");
        else
            warningDecorator.setMessage("No data");
        //Build decorator chain with a centred message
        decorator = warningDecorator;
    }
    else //Build the decorator chain based on the
         //current display mode
    {
        if (graphOptions.getDisplayMode()==
                            graphOptions.getCROSSHAIR())
```

```
            //decorator chain with X and Y axis,
              cross-hair and waveform
            decorator = concatenateDecorators(
                    waveformDecorator,
                    labelsDecorator,
                    crossHairDecorator,
                    axisYDecorator,
                    axisXDecorator);
        if (graphOptions.getDisplayMode()==
                                    graphOptions.getZOOM())
            //decorator chain with X and Y axis,
            //zoom box and waveform
            decorator = concatenateDecorators(
                    waveformDecorator,
                    labelsDecorator,
                    zoomDecorator,
                    axisYDecorator,
                    axisXDecorator);
        if (graphOptions.getDisplayMode()==
                                    graphOptions.getGRAB())
            //decorator chain with X and Y axis and waveform
            decorator = concatenateDecorators(
                    waveformDecorator,
                    labelsDecorator,
                    axisYDecorator,
                    axisXDecorator);
    }
}
/** Forms decorator chain. Note variable argument list. */
private GraphDecorator concatenateDecorators
                            (GraphDecorator ... decorators)
{
    for (int i = decorators.length - 2; i >= 0; i--)
        decorators[i].setDecorator(decorators[i + 1]);
    return decorators[0];
}
public GraphDecorator getDecorator()
{
    return decorator; // move to accessor methods?
}
```

A side effect of the introduction of the `GraphMediator` class is that the calls to the various draw methods can also be made simpler because the `GraphMetrics` object no longer needs to be passed to them.

11.2.3 Our Final Product

A class diagram for the `EScope4` graphics domain is shown in Fig. 11.2. It can be seen that the use of the decorator pattern and the conversion of the

11.2 EScope4: Adding Zoom and Grab Options Using the Decorator Pattern

`GraphMetrics` class into a mediator have tidied up this domain. The facade class just maintains links to the mediator class and to the `GraphMouseHandler` inner class. The mediator maintains links to all of the other classes in the domain and is responsible for passing data from the graph options and graph data objects, for performing the coordinate transformations, and for assembling the decorator chain. So, even though this domain contains more classes and additional functionality, it looks better organized than Fig. 10.2.

The power of the decorator pattern, can be demonstrated by a tiny experiment which the reader can perform using the version of `Escope4` on the CD: If the `DrawCrossHair` class is added to the grab-mode decorator in the `GraphMediator` class, then a cross hair will be drawn over the graph as you would expect. But if you move this cross hair into the waveform and then change to grab mode, then it will remain stationary as the graph waveform is dragged over it. This is a new, interactive diagnostic and it has been created by the addition of one line of code to the decorator chain!

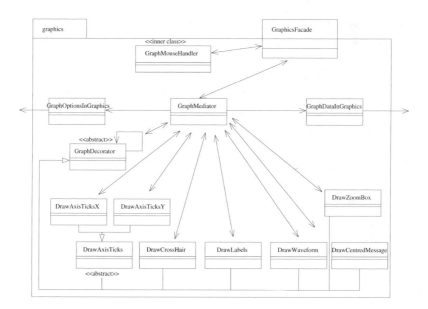

Fig. 11.2. Class diagram for the graphics domain for `EScope4`.

The value of this refactoring step can also be seen by looking at the `paintComponent` method of the `GraphicsFacade`. This method is now quite clear and uncluttered:

```
public void paintComponent(Graphics g)
{
    // Cast the graphics object to Graph2D
    Graphics2D g2 = (Graphics2D) g;
```

```
// Get plot size
Dimension size = getSize();
int xSize = size.width;
int ySize = size.height;
// find border size
int borderSize;
if (med.getGraphOptions().getDefaultBorder())
{
    borderSize = calcBorderSize(g2);
    med.getGraphOptions().setTickPix(4);
}
else
{
    borderSize = 0;
    med.getGraphOptions().setTickPix(0);
}
// Set graph and border size in med
med.setXSize(xSize);
med.setYSize(ySize);
med.setBorderSize(borderSize);
// Set background Color
g2.setColor(med.getGraphOptions().selectColor(
    med.getGraphOptions().getGraphColor()));
g2.fill(new Rectangle2D.Double(0, 0, xSize, ySize));
//call draw method of the decorator chain
med.getDecorator().draw(g2);
}
```

12
Patterns at Work: Multiple Waves

In the next two chapters, we will not introduce any new patterns. Instead, we will extend the graphics capability of EScope to include multiple waveforms (here in Chapter 12) and multiple plots (in Chapter 13). We will also incorporate some interesting algorithms into our graphics domain. By the time we finish we will have a pretty sophisticated graphics display which will have some amazing interactivity.

As we improve the software in this way we hope to demonstrate the usefulness of patterns by showing that extending EScope to incorporate these additional requirements is a reasonably logical process which involves only localized changes to the code. When we need to, we will implement the changes with some anticipation of what might be further requirements in the future. This evolutionary process of software improvement often occurs in real life.

12.1 EScope5: Multiple Waveforms

In the EScope versions we have constructed so far, only one waveform can be displayed at a time. The possibility of displaying multiple waveforms in the one graph panel could be desirable since scientists are often interested in comparing one physical measurement against another. For example, in the h1data dataset it might make sense to plot the radio-frequency diagnostic signals, .rf:A14_4:input_2, .rf:A14_4:input_3 and .rf:A14_4:input_4, on the same graph. In this case it would also be desirable to be able to explore these multiple waveforms interactively, to, for example, read off values using the cross-hair diagnostic.

Let us first start with a description of some informal requirements for displaying a set of waveforms on the same panel.

12.1.1 Requirements

Our informal requirements are described by the following list.

1. One graph panel will be able to display multiple waveforms.
2. If datasets are selected by clicking the corresponding items in the file tree while holding down either the *Shift* or the *Ctrl* key, then their waveforms will be plotted on top of each other on the one graph. Holding down *Shift* will cause a range of waveforms to be selected. Holding down *Crtl* will select the waveforms one by one. (This convention is the way that items are selected from a file list under Windows.)
3. A different color can, optionally, be associated with each waveform. The colors will be specified in the graph options dialog.
4. We shall limit the number of colors (but not the number of waveforms) to three. (If there are more than three waveforms then the waveform at index i can be plotted using a color at an index of i modulo 3.)
5. The zoom and grab graph-manipulation options will work also with multiple waveforms. They will act on the entire collection of waveforms.
6. The cross-hair decorator will be changed so that when the mouse is pressed in the panel, the cross-hair will be attached to the *closest* waveform. The cross-hair will move along this same waveform until the mouse is released. The color of the cross-hair will be the same as the selected waveform.
7. The waveforms will be scaled so that they all fit in the panel at the same time.

At a first glance, these changes might require a deep re-arrangement of code. This would almost certainly have been the case if we had stuck with the unstructured `PreEscope` versions of our software, but the refactoring into domains and the use of the facade, adapter, template, mediator and decorator patterns will allow us to keep changes to small number of classes. An appropriate rule for good code organization might be expressed by the following statement:

> If a new feature is required, then it should be possible to express this feature as a set of requirements such that every requirement can be satisfied by a change in a small number of classes.

Every experienced programmer knows the frustration of fixing the "casualties of change" in poorly-structured software. Often, these side-effects are hard to diagnose and harder to fix since they affect portions of code which are, apparently, not related to the changes. Our aim is to minimize these frustrations as a result of good software organization.

12.1.2 Interfaces and External Requirements

Our articulated facade pattern has resulted in a good separation between classes in the three domains of `EScope`. This architecture also suggests a roadmap for modifying the codes:

1. Change the shared interfaces to account for the "external" requirements which communicate between packages.

2. Modify the implementing classes so that they implement these external requirements. In the case of facade interfaces, this step will be performed for each package in isolation.
3. Where these changes interact with another external requirement or an "internal" requirement, then schedule those requirements to be implemented next. Test where possible.
4. Finally, account for the remaining internal requirements which are local to a particular package.

Requirements 1 and 3 can be reflected in shared interfaces as follows:

1. Change the interface method `setGraphData()` of `GraphicsFacadeInterface` to pass an *array* of several `GraphDataInterface` instances instead of a single one. (This will need a corresponding implementation of the `setGraphData()` method in the graphics domain.)
2. Change the interface method `getLineColor()` of `GraphOptionsInterface` in order to accept an integer argument specifying the color index. (This will need a change to the corresponding `setLineColor` method of `GraphOptions` and both methods will be implemented in the GUI domain.)

Of the remaining requirements, requirement 2 will map to changes within the GUI domain and requirements 4-7 will map to classes within the graphics domain.

12.1.3 Plotting an Array of Waveforms

One of our first design decisions has already been made for us when we said that we expect an *array* of waveform data to be passed from the GUI to the graphics domain. Having specified this particular data-structure, we need to place appropriate code into the graphics domain to handle it. In keeping with the existing structure of the code, we would expect this array to be passed to the graphics facade where it would be converted to an array of `GraphDataInGraphics` objects which would then be passed to the `GraphMediator`.

A decision now needs to be taken about how much to modify the `DrawWaveform` class. Theoretically, the flexibility of the decorator pattern allows us to chain together several `DrawWaveform` objects and this will prepare us for any future requirements which involve interaction with individual waveforms (such as dragging one waveform only). We need to make a minor modification to the `DrawWaveform` class in order to include an index of the particular dataset being plotted. Then, in the `createDecorators` method of `GraphMediator` (see Appendix F.1 on page 351), we concatenate the `DrawWaveform` objects to the decorator chain after setting up the axes, labels and mouse decorators.

Another policy decision needs to be made about which waveform plot should be used for the axis labels. We choose to use the first plot in the array.

A side-effect of this decision is that the `GraphMediator` class must be able to return a single waveform dataset for setting the axis labels. (Of course, it must also return the complete array of waveforms for plotting.)

A further policy decision needs to be taken about the treatment of invalid or non-existent graph data. We choose to set the bad-data flags whenever any one of the group of waveform datasets is bad. An alternative treatment would be to exclude this dataset from the group and print a diagnostic message.

Finally, we need to decide on the overall transformation parameters. Requirement 7 says that we want to fit all of the data into the plot. This means that the `update` method of the `GraphMediator` should be modified to step through the group of waveforms and find the global maximum and minimum data extents in the X and Y directions. These values will then be used to define the graph transformations.

After a consideration of the above issues, requirement 1 could be and implemented and tested. The testing could use hard-coded calls from the GUI facade to plot several datasets at once.

Requirements 3 and 4, for setting the colors of individual waveforms, could be scheduled for implementation at this time. They will need additional options for choosing waveform colors to be implemented in `GraphOptionsDialog`. The associated color indices need to be stored in the `GraphOptions` object and accessed from the `GraphOptionsInGraphics` object. These are all straightforward modifications. The results can be tested in the same way as for requirement 1.

Now for a bonus. Requirement 5, to do with the zoom and grab mouse operations, should just work fine without any code modifications. Seldom is life this easy! The cross-hair diagnostic will not be nearly so easy and will be discussed in Section 12.1.5 below.

12.1.4 Modifications to the GUI Domain

The implementation of requirement 2, for selecting multiple datasets from the file tree, is restricted to the method `valueChanged()` of the inner `DataTreeSelectionListener` class of the `GuiFacade`. When handling multiple selections, this method retrieves a set of selections instead of a single one, and builds an array of `GraphData` objects. This part of the implementation would be quite straight-forward except that Swing does not directly return the currently selected items. Instead, the `TreeSelectionEvent.getPaths()` method returns the incremental array of `TreePath` objects which have been both *added* and *removed* in the present selection and it has a method `isAdded(pathIdx)` which will query whether a particular path has been added. So, unfortunately, we need to maintain our own record of "presently selected" objects and update this record with the incremental information.

Using an hash table, provided by the `java.util.Hashtable()` class, we can associate a collection of graph data objects with keys based on their file

12.1 EScope5: Multiple Waveforms

path names. Hash tables are efficient, random-access data structures whose keys are hashed up to fit into one of a number of "bins".

We can then add and delete elements of this hashed collection using the information supplied by the `getPaths()` method of `TreeSelectionEvent`. The coding for this is shown below. Although it is more complicated than we would have liked, it is well localized within the `GuiFacade` class. The only other change needed to this class is to declare the hash table as an instance field.

Once these changes have been incorporated into the GUI facade, the entire block of changes can be tested by selecting groups of files and having them plotted. The only remaining requirement is the behavior of the cross-hair diagnostic.

```
public void valueChanged( TreeSelectionEvent e )
{    // Get the new selection and the
    //corresponding data for each path
    TreePath[] paths = e.getPaths(); //'paths' contains
                        //selected and removed paths
    String shortFileAddresses = "";
    //Find path names and update hash table
    // of selected paths
    for(int pathIdx = 0; pathIdx < paths.length;
                                        pathIdx++)
    {
        Object[] path = paths[pathIdx].getPath();
        String shortFileAddress="";
        for ( int i = 1; i < path.length; i++ )
        {
            shortFileAddress = shortFileAddress +
                            path[i].toString();
        }
        // update hash table
        if (shortFileAddress != "")
                        // Do not plot "TOP" node
        {
            if(!e.isAddedPath(pathIdx))//If the path has
                                //been removed..
                treePathHash.remove(shortFileAddress);
                    //Remove from the hash table
            else
            { //new paths selected
                GraphDataInterface currData = (database.
                    getPlotData(shortFileAddress));
                treePathHash.put
                        (shortFileAddress, currData);
                            //insert it
            }
        }
    }
```

```
}
//The hash table is now contains
//waveforms to be displayed
int numGraphs = treePathHash.size();
GraphDataInterface[] graphs =
                new GraphDataInterface[numGraphs];
Enumeration graphEnum = treePathHash.elements();
                                            //waveforms
Enumeration pathEnum = treePathHash.keys();
                                            //path names
shortFileAddresses = "";
int graphIdx = 0;
while (graphEnum.hasMoreElements())
{
    graphs[graphIdx] = (GraphDataInterface)
                                graphEnum.nextElement();
    shortFileAddresses +=
            (String)pathEnum.nextElement() + " ";
    graphIdx++;
}
if (numGraphs > 0)
{
    GuiFacade.this.setDescription(shortFileAddresses);
    GuiFacade.this.graph.setGraphData(graphs);
}
GuiFacade.this.graph.repaint();  //no data
}
```

12.1.5 Drawing the Cross-Hair

In implementing this final requirement, we are faced with a similar decision to the one we made when drawing multiple waveforms: the decorator pattern enables us to define separate cross-hairs for each waveform and then to plot the one of these which is closest to the position of the mouse. An alternative strategy it to stick with only one `DrawCrossHair` object but to augment it to store an index for the "closest" waveform. We elect to follow this second strategy because it seems wasteful to create cross-hair objects and then not use them. (If multiple cross-hairs were really required on the graph then they could be created and chained together in the same way as the waveform decorators.)

Only a very minor change to the `DrawCrossHair` class is needed to retrieve an index of the closest waveform and then to use this index to retrieve the correct waveform, and correct color, needed to draw the cross-hair.

Actually finding the closest waveform to plot involves some more work and this can be split between the mediator and the `GraphDataInGraphics` class. The graph mediator class can have a method to step through the group of waveforms and find the one which is closest to the mouse. Once

the box itself would be drawn segment by segment and the waveforms would be copied directly from the off-line image buffer. If a group of waveforms were being grabbed and moved, then the off-line image would be rotated or translated appropriately and then copied.

Drawing to an off-line image buffer in Java is exactly the same as drawing to a window, provided the graphics context passed to the drawing methods comes from the image object. Thanks to this fact, and to our implementation of the decorator pattern, almost all of the changes needed to implement this optimization will be located in the `GraphicsFacade` class. The only other tiny modification needed is to return two decorators from `GraphMediator`: one for the waveforms and one for the rest of the graph.

The `GraphicsFacade` declarations needed to set up an image buffer can be seen in Appendix F.2.2 on page 355. The important declaration is that for `BufferedImage`. Other "affine transformation" declarations are needed to handle the transformation of the image while dragging:

```
// The following instances are used to draw
// onto the image buffer
private BufferedImage wavesImage;
                                //Offline waveform image
//Current Transformation for buffered image
private AffineTransform currWaveTrans =
                                new AffineTransform();
//Base Transform, computed every time
//the offline image is created
private AffineTransform baseWaveTrans;
```

The logic of using the image buffer can be seen in the `paintComponent` excerpt in Appendix F.2.2, on page 356, and the actual drawing of the waves onto this buffer can be seen in Appendix F.2.2 on page 357. The image buffer is defined to have a larger size than the actual displayed buffer (we set it to be 3 times as large in each of the X and Y directions) so that the central part can be grabbed and dragged. Apart from this, and some book-keeping work to ensure that the image translations are computed correctly, drawing onto the image buffer is the same as drawing onto a normal graphics panel: the background is filled in and the draw method of the waveform decorator is called.

Finally, when in grab mode, the `AffineTransform` object used to drag an image needs to be set after every mouse drag event. This is done in the mouse handler class shown in Appendix F.2.2 on page 359.

13.3 Our Final Product

Key parts of the listing of `EScope6` are shown in Appendix F.2. The full version is, of course, available on the CD. When playing with it, notice that the three mouse manipulation functions (grab, zoom and cross-hair) apply

13 Patterns at Work: Multiple Graphs

independently to each graphics window. You can try placing waveforms in several windows and selectively zooming one, grabbing another and so on. Notice that you can apply the mouse tools individually to every window, not only the selected window. In the next chapter, we will learn how we can synchronize the behavior of the cross-hair diagnostics across graph windows.

Figure 13.2 shows the GUI domain for `EScope6`. Apart from the new layout manager class, the only real change is that there is a creation arrow pointing from the `GuiFacade` to the `GraphicsFacade` class. The class structure of the rest of the software system, including the graphics domain, is unchanged.

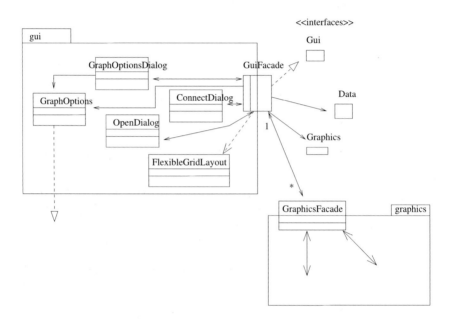

Fig. 13.2. Class diagram for the GUI domain for `EScope6`.

We think that you will really enjoy playing with this version of `EScope` together with the sample datasets supplied on the CD. But if you can wait until the next chapter, it will get even better!

14
Observer

The observer pattern is very widely used and it is subject to a wide range of terminology. Depending on which authors you read, the pattern relates, on the one hand, "sources" or "subjects" or "servers" or "managers" or "publishers" to, on the other hand, "observers" or "listeners" or "receivers" or "subscribers".

A version of the observer class diagram is shown in Fig. 14.1. The two interfaces are quite simple: The observer interface advertises, to the source, the name of the method which will be called whenever observer objects need to be notified of an event. On the source side, the interface advertises, to the observers, how to register as an observer, how to deregister as an observer and, optionally, the name of a method which can be called on the source object to force it to send a notification event to all observers.

The observer pattern allows for changes in an object's state to be broadcast to a collection of registered observers. It is sometimes called a "server push" or "publish-subscribe" pattern because the source object pushes, or publishes, information to the subscribers when there is a need to.

14.1 Pattern Description

We have already met the observer pattern when we attached "listeners" to GUI components in Section 3.4. In those examples, we typically wrote an inner class which implemented a particular listener interface. We then attached an instance of that inner class to the GUI component itself using a call such as `addActionListener(..)`. The GUI component object knows that registered listeners implement a particular method such as

```
public void actionPerformed(ActionEvent event)
```

The component can call this method and pass information to the listener through (in this example) an `ActionEvent` object. The listener object knows that the GUI component has a method to add listeners. This observer pattern is a quite restricted form of association between two classes.

In our treatment of trees in Swing, in Section 6.4, we saw how the observer pattern might be implemented inside the, "source", MDSTreeModel class of Sect. 6.5. In this class, a data-structure, EventListenerList, has links to a number of observer objects. The fireChangedEvent method can be used to broadcast a message to these observers. (One of the observer objects can be the JTree component.)

The observer pattern is useful for a variety of applications. Although they all implement the same interface, the observers respond in individual ways that are meaningful to them. The pattern is is relatively simple and it has the effect of decoupling the interfaces of the source class from the observer class. All that the observer class needs to know is that the source class implements its advertised interface. All that the source class needs to know is that it has to notify a number of objects which implement a listener interface. This type of interaction is so restricted that we consider it safe for use between classes in different domains of a software system.

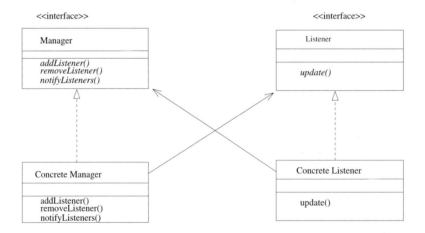

Fig. 14.1. An observer pattern.

In this chapter we will adopt a slightly pedagogical approach where we will implement the observer pattern in a more elaborate way than the revision of the software really needs but which will, hopefully, illustrate some of the subtleties of the pattern.

14.2 EScope7: Integrating Synchronized Interaction in Multiple Windows

Scientists are often interested in using graphical tools to explore relationships between different datasets. In the context of EScope, plotting waveforms on top of each other is one way of exploring their relationship. Another extremely useful feature is synchronized cross-hair dragging where the cross-hair in every waveform window can be dragged by just dragging in one of the windows. This enables a scientist to see what was happening in a particular diagnostic channel at exactly the same time that something else was happening in another channel. For this synchronized diagnostic to be effective, it might occasionally be useful to be able to set the same scales for each of the displayed waveform windows. We shall implement these synchronized diagnostics in this revision of the code.

Our synchronized interaction will be described by the following, informal requirements:

1. In cross-hair mode, dragging the cross-hair over one waveform window will cause the cross-hair in every other window be moved in a synchronized way.
2. The pop-up menu will be expanded to include the following items:
   All same X scale: Set the X scale of every waveform window to be the same as the current window.
   All same scale: Set the X and Y scale of every waveform to be the same as the current window.

These requirements will drive us to implement two observer patterns: a "points observer" which will communicate the position of the cross-hair points to the other graph windows (as well as to the GUI) and a "scale observer" which will communicate rescaling information from one graph window to all of the other graph windows.

The "manager" in both the point observer pattern and the scale observer pattern will be the `GraphicsFacade` class. The observers will be the `GuiFacade` and the `GraphicsFacade` classes, in the case of the points observer pattern, and the `GraphicsFacade` classes in the case of the scale observer pattern.

14.2.1 `sharedObserverInterfaces` Completes the Articulated Facade

We have noted that the GUI facade is already an observer of the cross-hair movement. In order to formalize this into an observer pattern, we need to move the `graphPointUpdated` method-signature from the `GuiFacadeInterface` to an observer interface. Because this interface will extend across domain boundaries, we need to place it into one of our existing interface packages or create a new package. We choose to create a new package called `sharedObserverInterfaces`. The reason for this is to separate the three types of inter-domain communication that we will allow in our, now complete, "articulated facade" pattern:

1. direct calls to facade interface methods, advertised as method signatures in the facade interfaces,
2. data channels of communication where certain data objects are advertised between domains, and
3. observer channels of communication which link source objects with observer objects between domains and which are advertised by their respective interfaces.

The directory structure of the software makes this point clear: there are 3 separate directories for each of the software domains and there are 3 separate directories for each type of interface. We think that the source-observer relationship is so restrictive that it can be safely be used to relate objects which are not facade objects. So we can tolerate some leakage from our nice domain structure providing the spirit of this pattern is adhered to. We first distinguish the "points" observer, which will communicate information across domain boundaries, from the "scale" observer which is totally confined to the graphics domain. The `GraphPointsListener` interface will be implemented by the `GuiFacade` to print out the data values under the cross-hair in a particular graph window. It will also be implemented by the `GraphicsFacade` to broadcast and update the cross-hair positions in all graph windows. Unfortunately, we have a name clash because `GraphicsFacade` already has a `graphPointUpdated` method which is used for something else. So we choose to change the name of this method in the GUI domain.

The source, `GraphPointsManager`, interface will be also located in the `sharedObserverInterfaces` domain and will be used by the `GuiFacade` to register (or "add" or "attach") itself to the source. Both this interface and the `GraphPointsListener` interface will be implemented by the `AbstractGraphicsFacade` class and, through it, the `GraphicsFacade`. Their listings follow:

```
package sharedObserverInterfaces;

public interface GraphPointsListener
{
    public void pointsUpdated(double xCrossHair,
                  double yCrossHair, boolean inRange);
}
```

```
package sharedObserverInterfaces;

public interface GraphPointsManager
{
    public void addGraphPointsListener
                            (GraphPointsListener l);
    public void removeGraphPointsListener
```

this is found, its index can be stored in the `GraphMediator` object and retrieved by the `DrawCrossHair` decorator object when the graph is drawn. The `recordClosestWave` method of `GraphMediator` is reasonably straightforward and is shown in Appendix F.1.1 on page 351. It steps through the graph data array and calls a `getYDistance(xVal,yVal)` method on each one.

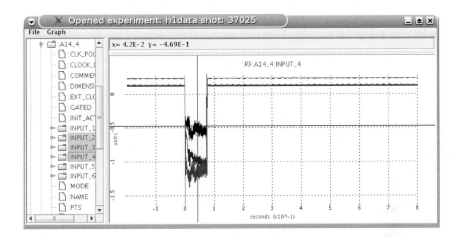

Fig. 12.1. A screen shot of `EScope5` showing multiple waveforms and the the cross-hair diagnostic.

Figure 12.1 shows a screen shot of `EScope5` in action with the cross-hair diagnostic. Because the cross-hair is an *interactive* diagnostic, it is possible to simplify the algorithm to compute the closest waveform. In particular, one can make the following simplifications.

- The closest waveform can be computed using distances in the Y (vertical) direction only. This simplification means that it is easier to search the waveform arrays to find the closest distance than it would be if 2D Euclidean distances were to be computed.
- There is no need to worry about the loss of data due to filtering of the waveform. That is, we do not need to search the data points which have been filtered out to find the closest waveform.

Our point about the diagnostic being *interactive* is the following: if the algorithm does not find an exact solution every time, the user only needs to move the mouse slightly to have it stick onto the appropriate waveform.

It is natural to locate the "closest distance" calculation in the same class as the data itself (`GraphDataInGraphics`). Its listing can be seen in Appendix F.1.2, on page 353, and it is appropriate to make some comments about its algorithm here:

1. Before being called from `GraphMediator` the coordinates of the mouse position are converted to (world) data coordinates.
2. The changes to `GraphDataInGraphics` are contained in the two methods `getYDistance` and `getXIdx`.
3. `getYDistance` retrieves the array of Y data from the graph dataset. It then calls `getXIdx` to compute the index which gives the closest X value to the X position of the mouse. This value is used to index the Y array and to compute the distance between the mouse Y and the value in the data array.
4. `getXIdx` retrieves the array of X data from the graph dataset. It is assumed that this array increases monotonically. Bisection is then used repeatedly over the interval of X values to find the data X value which is closest to the X coordinate of the mouse
5. Because a converted pixel coordinate is being compared with a full X data array, the algorithm will possibly return one end of an interval of possible index values. As discussed above, this is not likely to be a problem because users will move the mouse slightly if it is not sticking on the correct waveform.

12.2 Our Final Product

Had it not been for the messiness forced on us by Swing, this revision of `EScope` would have been quite intuitive with the most demanding part being the construction of an algorithm to find the closest waveform to a mouse click. As it is, the changes have been very well localized due to our software architecture. It is worth-while to summarize them again here:

1. We store an array of `GraphDataInGraphics` objects in `GraphMediator`.
2. We modify `DrawWaveform` so that its constructor gets passed the index of the waveform dataset to be plotted. We then modify the construction of the decorator chain so that multiple waveform decorators are constructed and chained together.
3. We modify `GuiFacade` to process multiple selections from the file tree. This is entirely contained in the `valueChanged` method which accepts a `TreeSelectionEvent` as its argument. A hash table needs to be used to properly process multiple selections.
4. We modify `DrawCrossHair` to access the index of the closest waveform in the graph data array. This is used to select the correct waveform for the cross-hair and to find a color index for the color of the cross-hair.
5. We add the method `recordClosestWave` to `GraphMediator` and the methods `getYDistance` and `getXIdx` to `GraphDataInGraphics`.
6. We ensure the changes integrate together properly. In particular, we ensure that the correct information is passed from the GUI to the graphics mediator via the graphics facade.

The class diagram for the graphics domain of ESCope5 is identical to that for ESCope4 with the exception of mulitplicity indicators and we shall not show it here. Overall, we are satisfied with the structure of this version of the code but observe that the redrawing of the graphs can be a somewhat slow – especially if a user is grabbing and dragging several waveforms at the same time. Using OpenGL hardware acceleration, as described in Sec. 5.2.1, will help to speed things up a bit. But another technique which we will describe in the following chapter will help a great deal. In the next chapter, we will also generalize the display to have multiple waveforms in multiple graphs.

13

Patterns at Work: Multiple Graphs

In this chapter we will enhance the graphics functionality of `EScope` to have multiple graphics windows with multiple waveforms in each. This modification is so suited to the architecture of `EScope` that it is almost trivial! We shall make it considerably less trivial by incorporating a custom layout-manager into the graphics panel which will provide asymmetric, resizable graphics windows with a minimum of wasted space. We shall also introduce a speed optimization which should fix up some of the poor performance which readers may have noticed with previous versions of `EScope`.

13.1 EScope6: Multiple Windows in EScope

`EScope5` is able to display one or more waveforms in a single window. The selection of the datasets to be displayed is achieved by clicking the corresponding icons in the file tree. If more than one window were displayed in the interface, it will be necessary to find a way of selecting a given window and specifying that it will receive waveforms. One possible way of doing this is to right-click a window to activate a pop-up menu which contains an item which allows that window to be selected. After being selected, some visual feedback, such as a a red border around the window, will signify that the window is active and ready to receive waveforms.

The changes in `EScope` to handle multiple windows might, therefore, be described by the following, informal requirements:

1. It will be possible to define a number of windows, possibly of different size, for waveform display.
2. The number and layout of the windows will be specified by defining the number of columns and, for each column, the number of rows. This means that the grid of windows can be asymmetric.
3. The definition of columns and rows layout is to be specified in a "property file" to be read during startup.

4. Initially, rows will have the same height within a given column. It is possible to dynamically resize the width of the columns and the height of the windows by dragging small "knob icons" which are displayed at the bottom of each window (except the lowermost ones) and at the right of each column (except the rightmost one). Clicking a knob with the middle mouse button will cause the window to revert to its default size.
5. The selection of the window to receive waveforms is done by clicking the right mouse button within the window to activate a pop-up menu. When the "select" menu item is chosen, the selected window is to be highlighted by a red border around it.

We have already discussed how a proper organization of software should localize the required changes for the integration of new requirements. In a well-written code, each new requirement should imply changes in a limited number of classes. If we look at the above requirements and, in particular, to requirement 4, we realize that the changes are not trivial. The dynamic resizing behavior is sophisticated and is not supported by any layout manager provided by Java. In fact, some "turgid programming" is required, but this will be limited to a single class which we will give to you and describe in general terms. The remaining changes will be limited to few lines in a couple of classes. Figure 13.1 shows a screen shot of this version of the software: `EScope6`.

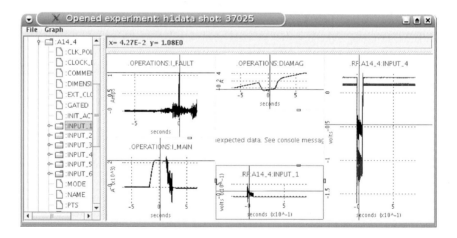

Fig. 13.1. A screen shot of `EScope6` showing multiple waveforms in multiple, asymmetric windows. The cross-hair diagnostic is displayed in each window.

13.1.1 Designing for Multiple Windows

Multiple windows will mean multiple graphics facades. There are a few design approaches to deal with this. The most attractive of these is to have multiple `GraphicsFacade` objects declared and managed by the `GuiFacade`. With this approach, the main graphics panel of the GUI would be a container and it would be filled with graph windows each of which would be associated with one `GraphicsFacade` object. This main graphics panel would need to have a layout manager – either one of the conventional Swing layouts (flow, border, grid, grid-bag and so on) or our custom layout. So it is fitting that our new layout-manager be located in the GUI domain.

If multiple graphics facades (representing multiple instances of the graphics domain) are to be declared by, and associated with, one `GuiFacade` object, then the `GuiFacade` will need to keep track of which `GraphicsFacade` is presently active. So if a new graph window is selected by the user, then the corresponding `GraphicsFacade` object needs to register itself with the `GuiFacade`. It turns out that the `GuiFacade` only needs to maintain a handle on this one, active `GraphicsFacade`. The other graphics windows are self-managing and this modification is a very elegant one indeed. The basic steps are:

1. The top-level, `EScope6` program remains unchanged. It creates the domains including one instance of the `GraphicsFacade` which gets passed to the `initialise` method of the `GuiFacade`.
2. The `GuiFacade` declares a new `JPanel` which is to be a container for the collection of graph windows. It associates a layout-manager with this panel and then it creates all of the other `GraphicsFacade` objects and adds every `GraphicsFacade` to the main panel. It retains a handle on only one `GraphicsFacade` which will be the default active window.
3. The `GuiFacade` needs to export the interface of two new methods: `getSelectedGraph` and `setSelectedGraph`. These methods will be called, once each, in the `GraphicsFacade` class; either to set a particular `GraphicsFacade` object as being selected (after being chosen by a user from the pop-up menu) or to test whether it is selected (and then draw a highlighted boundary). The two new methods need to be implemented in the `GuiFacade` class.
4. The only other coding needed is to include a pop-up menu with a "select" menu item in the `GraphicsFacade` class.

In summary, a minimum of changes is needed to two classes together with the advertising of two new methods in the `GuiFacadeInterface`. It is, apparently, that simple.

A Factory Method for the `GraphicsFacade` Class

There is a subtle issue with the apparent simplicity of the procedure outlined in the previous section. If the `GuiFacade` object is to create a collection of

GraphicsFacade objects and then add them to a graphics panel, then it would need to import the graphics domain in order to call the GraphicsFacade constructor. This is because Java does not export constructors in interfaces. But this is a problem, because we have worked very hard to ensure that the GuiFacade class only has access to the interface for the GraphicsFacade (via the AbstractGraphicsFacade abstract class in the sharedInterfaces package) and only the main program in our system is meant to have import access to all of the domains.

This subtle problem can be solved by using a factory method in the GraphicsFacade class. As explained in Chapter 17, factory methods are an alternative to constructors for constructing objects. They have several advantages. One is that the signature of a factory method can be exported in an Java interface whereas that of a constructor can not. For this reason, we include a new method, getGraphicsFacade in the GraphicsFacade class:

```
public AbstractGraphicsFacade getGraphicsFacade()
{
    GraphicsFacade newGraph = new GraphicsFacade();
    return newGraph;
}
```

It can be seen that all this method does is to call the constructor, but it returns the object as one of type AbstractGraphicsFacade so it can be called by another class which only has access to the sharedInterfaces package (and no access to the graphicsDomain). But the calling object needs to access a GraphicsFacade object in order to invoke this method! Is this a problem? No problem, because the GuiFacade object is initialized by being passed a GraphicsFacade object from the main, EScope program.

Reading Parameters from a Property File

A final problem for the revised GuiFacade class is how to specify the number, and grid geometry, of the GraphicsFacade windows. These could be hard-coded in order to test the operation of the software, but a better solution is needed for production use.

"Property files" are text files which contain string identifiers (the parameter names) and parameters. They can be conveniently loaded and read using the java.util.Properties class as shown in the listing of Appendix F.2.1 on page 354. This useful class inherits from java.util.Hashtable and it uses a hash table to store the property data based on a hash key of its (String) name. (Very usefully, the Properties class also contains methods to read and write XML data but this, unfortunately, is outside the scope of this book.) The following example shows a property file which specifies a grid of graph windows which has three columns with two, three and one rows, respectively:

```
num_columns=3
column_1.num_rows = 2
column_2.num_rows = 3
column_3.num_rows = 1
```

13.1.2 The Flexible Grid Layout Manager

This complicated piece of code is contained in a single class, `Flexible-GridLayout`, which implements the `java.awt.LayoutManager` interface. It is so complicated that we do not include a listing in the appendices and interested readers are referred to the accompanying CD.

Although complicated, this class demonstrates an advantage of the graphical class management carried out by Java's Abstract Window Toolkit which defines a clear boundary between graphical management and the computations required for a proper component layout in a container. In fact, laying out components within the container window, with dynamic resizing using the mouse, requires interactions between the container, the contained elements and the knob components. Such a situation could easily lead to a messy implementation where the dependencies among components would make the code hard to understand and maintain.

Interestingly, the `FlexibleGridLayout` class does not explicitly draw the knob shapes onto the container. Instead, the knobs are defined, in an inner class, to be components in their own right and they are managed by the outer, `FlexibleGridLayout`, class. This means that the knobs have their own `paint` method and they can respond to mouse events.

`FlexibleGridLayout` implements the `java.awt.LayoutManager` interface and its methods include `addComponent()` and `layoutContainer()`. The `addComponent()` method is called to inform the layout manager when a new component has been added to the container. The `layoutContainer()` method is expected to perform the sizing and positioning of all of the graph panels. For each graph panel, a data structure is maintained to hold the component reference, the corresponding knob reference, and the percentage of the column height assigned to that component. This information is defined by the `ComponentInfo` inner class of `FlexibleGridLayout`. For each column, `FlexibleGridLayout` also maintains the corresponding knob reference and the percentage of the container width assigned to that column.

Based on all this information, `layoutContainer()` resizes every graph panel and positions the additional knob components. This information is updated when a knob is dragged. For example, if a bottom knob is dragged the height of the corresponding graph panel is changed according to the amount of mouse movement in the vertical direction. The heights of all the other components in the same column are also adjusted with the constraint that the height of every graph panel cannot fall below a given threshold. Similarly, when a side knob is dragged, the corresponding column widths are modified so long that the width of every column does not fall below a given threshold.

In summary, despite the complexity of the `FlexibleGridLayout` algorithm, all the required code is contained within a single class. The other `EScope` classes are unaffected except for `GuiFacade` which needs to declare an instance of `FlexibleGridLayout` as its layout manager.

`FlexibleGridLayout` is self-contained and can be used every time graphical components need to be laid out in a container. We hope that you will enjoy using it and that you will find a "reuse" for it in your own software!

13.2 Image Buffering: A Useful Graphics Trick

Some readers will have noticed that multiple waveforms in `EScope5` can be annoyingly slow to interact with. This is particularly the case if there are many, zoomed waveforms and you are tying to perform an operation such as grabbing and dragging. The reason for this slowness is the number of operations which must be performed for every repaint of the graphics display: the entire set of waveform coordinates needs to be calculated and each tiny `Line2D` wave segment must be drawn individually.

At present, our graphics display is entirely repainted if there is some change to it. Obviously, this needs to be done in some situations, such as

- changing the size of the graphics window,
- setting new graph data or graph options,
- performing a zoom operation,

where the appearance of the waveforms will change on repaint. But there are three important situations where a complete repaint is not necessary:

1. Dragging a cross-hair has the visual effect of just changing the cross-hair position while the background is left unchanged.
2. Drawing a zoom box by dragging the mouse just draws a box on an unchanged background of waveforms.
3. Translating the waveforms by using the grab tool and dragging the mouse will change their appearance, but this change is just a rigid shift of the waveforms. It should be possible to find a more efficient way of repainting them in this case.

All of these situations have something in common: They occur when the mouse is *in the process of being dragged*. So if we distinguish the "dragging" mouse events from the "just pressed" or "just released" mouse events then we can isolate the situations where the background waveforms will either not change at all (cross-hair and zoom tools) or change only by a rigid translation (grab tool).

An extremely useful technique in computer graphics is the idea of "image buffering". In this technique a graphical image, or part of a graphical image, is painted onto a buffer which is different to the one being displayed on the device. When it comes to display the true image, then the off-line buffer is copied directly to the display. This image copy operation is extremely fast – much faster than drawing every line segment in a group of waveforms. Image buffering is useful when part of the image does not need to be redrawn every time. So, if a zoom box is to be drawn on top of a group of waveforms then

14.2 EScope7: Integrating Synchronized Interaction in Multiple Windows

```
                                        ( GraphPointsListener  l );

    //Notification of points listeners
    public void notifyGraphPointsUpdated(double x, double y,
                                         boolean inRange);
}
```

The implementation of the `GraphPointsListener` by the `GuiFacade` class only requires a method name change but the `GuiFacade` will need to register itself by calling the `addGraphPointsListener` methods of the `GraphicsFacade` objects as it creates them.

14.2.2 Management of a Collection of Graphics Facades

Within the `GraphicsFacade` things are more complicated. Because we need to synchronize across graph windows we need to have a mechanism for one window to know about all of the others. This can be achieved by using a *static* data-structure. We modify `GraphicsFacade` to declare an `ArrayList` of `GraphicsFacade` objects, an `ArrayList` of `GraphPointsListeners` and an `ArrayList` of `GraphScaleListeners`. When each `GraphicsFacade` object is constructed, it needs to be added to the appropriate, static data-structure:

```java
public class GraphicsFacade extends AbstractGraphicsFacade
             implements GraphScaleManager, GraphScaleListener
{
    private static ArrayList<GraphicsFacade> graphList=null;
    ...
    //ArrayList to store the points listeners.
    private static ArrayList<GraphPointsListener>
                   pointsListeners = new ArrayList();
    //ArrayList to store the scale listeners.
    private static ArrayList<GraphScaleListener>
                   scaleListeners = new ArrayList();
    ......
    public GraphicsFacade()
    {
        med = new GraphMediator(this);
        addMouseListener(mouseHandler);
        addMouseMotionListener(mouseHandler);
        graphList.add(this);
        addGraphPointsListener(this);
        addGraphScaleListener(this);
        createPopupMenu();
    }
    public AbstractGraphicsFacade getGraphicsFacade()
    {
        GraphicsFacade newGraph = new GraphicsFacade();
        return newGraph;
    }
```

It might be reasonably argued that having three static collections to manage the graph windows, themselves, as well as the two observer patterns is much too elaborate. But it is the most general way of self-managing the collection of `GraphicsFacade` objects:

- The `graphList` list is used to contact each `graphicsFacade` object to, for example, update graphics options.
- The `pointsListeners` list is used to implement the points observer pattern. It includes the `GuiFacade` as well as each of the graph windows.
- The `scaleListeners` list is used to implement the scale observer pattern. Although this listener list is the same as `graphList`, this might not always be the case. For example, a future modification of the code might allow graph windows to be individually added and removed from this list. In this way individual windows could have the same scale as some, but not all, of the other windows.

A broadcast update of graph options can be implemented as follows:

```java
/** Sets graph options locally and broadcasts them
    to other windows.*/
public void setGraphOptions
                    (GraphOptionsInterface graphOpt)
{
    GraphOptionsInGraphics options =
                    new GraphOptionsInGraphics(graphOpt);
    med.updateOptions(options);
    //Set identical graph options for other graphs
    for (int i=0; i<graphList.size(); i++)
        if (graphList.get(i)!=this)
            graphList.get(i).
                            setThisGraphOptions(options);
}

/** Set broadcast graph options locally */
public void setThisGraphOptions
                    (GraphOptionsInGraphics options)
{
    med.updateOptions(options);
}
```

The implementation of many of the observer interface methods in the `GraphicsFacade` class is quite straight-forward. For example, the following excerpt shows the implementation of the `notifyGraphPointsUpdated` method:

```java
public void notifyGraphPointsUpdated(double x, double y,
                                     boolean inRange)
{
    for(int i = 0; i < pointsListeners.size(); i++)
    {
```

```
            GraphPointsListener currListener =
                (GraphPointsListener) pointsListeners.get(i);
            currListener.pointsUpdated(x, y, inRange);
        }
    }
```

The implementation of the `pointsUpdated` method shows that a particular graph window first checks to see whether it is being dragged and, if not, it then updates the X coordinate of the cross-hair and repaints itself:

```
    //Implementation of GraphPointsListener interface
    //Move cross-hair in this window to be at the same X as
    //active cross-hair
    public void pointsUpdated(double x, double y,
                                              boolean inRange)
    {
        if(mouseState == DRAGGING) return;
                           //this is the active window
        med.setCurrX(med.getXPixel(x));
        repaint();
    }
```

14.2.3 The Graph Scale Interfaces

Our requirements specify that we need to implement two new pop-up dialog options to broadcast scaling information from one graph window to the group. It makes sense for the graphs to have uniform scales for the X (time) axis when using the synchronized cross-hair diagnostic. There are also situations where users would want to easily compare the extent of features in the Y direction. So we will include options to synchronize only the X axes and to synchronize both axes.

The source of these events will, once again, be a particular `GuiFacade` object and the observers will be the other `GuiFacade` objects. Because all of the affected classes lie in the graphics domain there is no need to add the interfaces to the `sharedObserverInterfaces` directory (or to even have interfaces, but we will do so to be true to the observer pattern!). So we define a `GraphScaleListener` and a `GraphScaleManager` interface and have the `GraphicsFacade` class implement both of them.

In order to synchronize scales, the maximum and minimum extents of the source waveform need to be broadcast to the mediator objects in the other graph windows. We can do this by defining a `GraphUpdateEvent` object which is a data packet which gets constructed and broadcast whenever either of the scale menu items are chosen. The listing for this event class, for the scale interfaces, for the complete pop-up menu and for the methods which call the scale update methods can be found in Appendix F.3 on page 360. The rest of the listener and manager methods are quite predictable and can be seen in the full `EScope7` code on the CD.

14.2.4 Our Final Product

At this point it is worth listing the changes we have made to implement synchronized interaction:

- We have established a `sharedObserverInterfaces` package and placed the `GraphPointsListener` and `GraphPointsManager` interfaces there. The `GuiFacade` now implements the listener interface and the `AbstractGraphicsFacade` implements both the listener and manager interfaces.
- `GuiFacade` has been modified to implement the `GraphPointsListener` interface. These changes are quite minimal (including renaming one method).
- The new scale interfaces have been placed in the graphics domain. The graphics facade implements both of them. We also define a `GraphUpdateEvent` class which is located in the graphics domain.
- The lion's share of the work is in modifications to the `GraphicsFacade` class. This class needs to define and maintain collections of listeners (and the other graph windows) and manage them appropriately. The general organization is that mouse events, or a menu selection, will cause a group of listeners to be notified.

Many of these modifications have involved the addition of new interfaces or of new methods. Many of the changes have been very minor and even the modifications to the `GraphicsFacade` class have been reasonably predictable once you have the idea of managing a static collection of objects of the same type as the enclosing class. The class diagram of `EScope7` is very similar to that for `EScope6` and the interested reader could sketch it: There will be some dashed (implementation) arrows from the `GuiFacade` and `GraphicsFacade` to the observer interfaces and the graphics domain has the new `GraphUpdateEvent` which is "used by" the facade.

We may have gone "a bit over-board" with this implementation of the observer pattern. There are many informal ways to have one object listening to another and we probably could have done without the explicit scale interfaces seeing as the `GraphicsFacade` class is only advertising them to itself.

This version of the software is great fun to play with! It is basically as good as it gets for the appearance of `EScope`. The following chapters will deal with improvements to some of the "behind the scenes" aspects of the code and we shall particularly be focusing on the data server domain.

15
Proxy

As its name implies, the proxy pattern is about one object standing in for another. The proxy pattern consists of a proxy class and a target class which both implement the same interface. It is common for the client to first connect to a proxy object which decides whether or not to contact the true target. A possible class diagram for the proxy pattern is shown in Fig. 15.1.

The proxy pattern is particularly used for *caching* of remote data and it is very relevant for networked e-Science applications like `EScope`. If a dataset is located on a remote `MDSPlus` server, then it makes a lot of sense to save it in a local cache for possible reuse rather than downloading it repeatedly.

The proxy pattern can also be used to implement a layer of *security* where the proxy object filters requests from the client to the target. This is also of relevance to Grid-enabled e-Science applications but it will not be discussed further here.

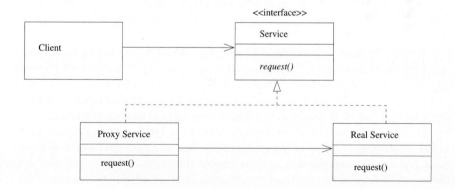

Fig. 15.1. Proxy pattern.

15.1 `EScope8`: Implementation of a Local Data Cache

We can express the requirements for data caching in `EScope` as follows:

1. The system has to provide local data caching for signals which are accessed remotely.
2. Each time a new signal is requested, a check is done to verify whether data are stored in a local file. If there is no local data, then the signal is retrieved from the remote server.
3. When a new signal has been received from the network, a copy is made in the local file space.

The above requirements define a very simple caching policy. In practice, more sophisticated caching procedures are often required such as:

1. Calculate the total size of files in the cache directory and purge older files if storage space is running out.
2. Implement persistent caching to allow cache files to be accessed in different sessions. This requires some sort of permanent indexing mechanism.
3. Implement multi-level caches. For example, a dataset could be stored in a linked list before, or as well as, being written to a local file. An example of a linked list cache was given in Section 6.1.

15.1.1 The `DataServerProxy` Class

The packaged structure of `EScope` makes it very easy to determine where a proxy data server should be implemented. In fact the proxy pattern can be implemented with only the addition of one new class which implements `DataServerFacadeInterface`. This new class, `DataServerProxy`, will be the new facade for the data server domain. It creates and maintains a link to the old `DataServerFacade` class and delegates almost all of its methods directly to it without modification.

In the previous versions of `EScope`, the `DataServerFacade` processed a `getPlotData` request by contacting the server and repeatedly downloading the data needed to build a `GraphData` object. Our `DataServerProxy` class can read and write these objects directly providing the `GraphData` class is *serialized*. That is, it needs to implement the `Serializable` interface.

Our implementation of the `DataServerProxy.getPlotData` method uses a list to store the names of the cached datasets. This list gets searched to see whether a requested dataset is there before the proxy contacts the real data server. A listing of this method can be seen in Appendix F.4.1 on page 363.

15.1.2 Our Final Product

This has been a very easy pattern to implement. A partial view of the `dataServerDomain` class diagram is shown in Fig. 15.2. Both the proxy and

15.1 EScope8: Implementation of a Local Data Cache   183

the real `DataServerFacade` class are drawn on the domain boundary as they both implement the facade interface.

This version of the code operates identically to `EScope7` apart from the local caching. You should notice the speed-up if you download a big dataset twice. Just to be sure that the cache is operating, we have included some diagnostic statements in the version of this code on the CD.

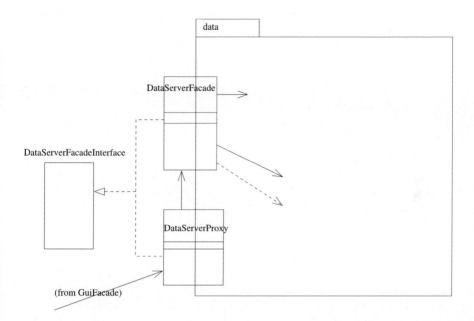

Fig. 15.2. Partial class diagram for the data server domain for `EScope8`. The real facade and the proxy facade sit on the boundary of this domain and implement the same facade interface.

16
State

16.1 Pattern Description

Objects are a combination of state (their data attributes) and behavior (their methods). The idea of an object being a "state machine" was central to the evolution of object-oriented programming. Active objects emerge naturally in simulations of engineering systems. Even in normal business applications, you tend to find objects which are state machines. In real-time systems you find them a lot.

The state pattern is a way of clearly coding the state behavior of a dynamic object. As generally implemented in Java, you define a class which contains inner classes which implement a well defined state interface. The outer class defines the *context* of the state machine object. It keeps a handle on the (inner) object which represents the current state. State transitions are managed from the methods of the inner objects themselves.

The state pattern can substitute for algorithms which have lots of conditional (`if .. then .. else`) statements and where the states themselves are stored as combinations of boolean flags. It can become quite difficult to debug algorithms of this nature and some people find that the state pattern is a much clearer alternative. Extending a state pattern is a matter of defining new, state inner-classes and then modifying the linkages between states. The state pattern translates well to diagrammatic representations of active objects ("state transition diagrams").

One version of a state pattern is shown in Fig. 16.1. In this figure, the concrete states are explicitly tagged as being inner classes of the "state manager" class. This is not essential for the pattern but it is very convenient as we will see below.

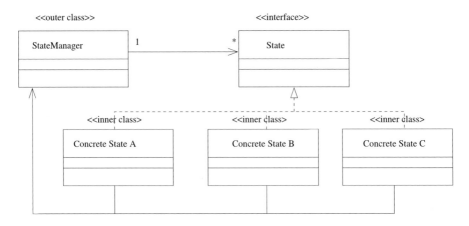

Fig. 16.1. State pattern.

16.2 Escope9: A State Pattern for the DataServerFacade

At present the `DataServerFacade` class implements the communication protocol with an `mdsip` data server. The following methods are defined in the `DataServerFacadeInterface` and their behavior is dependent on the state of the server connection:

- `connect`: establishes a network connection with the data server;
- `disconnect`: disconnects from a data server;
- `open`: opens an experiment database assuming a connection has been made already;
- `close`: closes the current experiment;
- `evaluate`: evaluates an `MDSplus` expression assuming that a valid experiment is open.

The usual way for handling system states is to use boolean flags, and this has been done in `DataServerFacade` using the variables `isConnected` and `isOpen`. Although this is might be satisfactory for a simple system, the number of boolean flags typically increases with the complexity of the dynamic object being modelled. It is, in fact, quite difficult to understand a code cluttered by `if` statements based on several different flags. Moreover, what typically happens when some new condition is added is that new flags are inserted in the code with potential side effects. Experienced programmers know the frustration of trying to understand such a code and getting lost in the huge number of possible conditions. (Recall in fact that with n flags, we face potentially 2^n different conditions.) This means that, in practice, many combinations of flag values are illegal and may lead to unexpected behavior. (Even with only two flags, we can have illegal states in the `DataServerFacade`: what is the meaning of `isConnected == false` and `isOpen == true`?)

16.2 Escope9: A State Pattern for the DataServerFacade

One solution to this problem is to get rid of every boolean flag and to use an integer variable to keep track of the current state. In this case it is possible to give a named constant to every state (for example, DISCONNECTED, CONNECTED, OPEN) and the current system status can be detected by testing a single variable, rather than a number of different flags. The if statements scattered in the code will be substituted by case statements which describe the state behavior more clearly. (We have partially adopted this approach in the use of a state variable, mouseState, in GraphicsFacade to store the status of mouse interaction with a graph window).

Another alternative is the state pattern which does away with all of the if and case statements. The State pattern defines an object instance for every state defined in the system. All the state objects implement a given interface (or extend an abstract class) defining the methods which depend on the current system state. An object reference will hold the instance describing the current state.

Integrating the State pattern into the DataServerFacade is simpler than one would expect and, rather than making radical changes in the existing implementation, it can be accomplished by *adding more code*. The required steps are summarized in the following sections.

16.2.1 Common Interface

A common interface, ServerState is defined for all state objects:

```
package dataServerDomain;
import java.io.*;

//State interface description. The interface will
//be implemented by inner classes of DataServerFacade.
interface ServerState
{
    void connect(String source) throws IOException;
    void disconnect() throws IOException;
    void open(String experiment, int shot)
                                throws IOException;
    void close() throws IOException;
    MDSDescriptor evaluate(String expression)
                                throws IOException;
    boolean isConnected();
    boolean isOpen();
}
```

16.2.2 State Inner Classes

We define the inner classes Disconnected, Connected and Opened. They implement the ServerState interface by, if appropriate, calling a method from the outer class to contact the MDSplus server. Note that the names of these outer methods have been changed because the `DataServerFacadeInterface` will now be implemented by state transitions as shown in the next section. We placed an "mds" prefix in front of the old names for the outer class methods.

A listing of the `Connected` state follows. Note that illegal transitions throw exceptions which need to be handled by the caller. This may or may not be what you really want to do; an alternative would be for illegal transitions to just do nothing.

```
//Class Connected state. Describes connection behavior when
//connected to a data server, but no experiment open.
private class ConnectedState implements ServerState
{
    public void connect(String source) throws IOException
    {
        throw new IOException("Already connected");
    }
    public void disconnect() throws IOException
    {
        mdsDisconnect();
    }
    public void open(String experiment, int shot)
                                          throws IOException
    {
        mdsOpen(experiment, shot);
    }
    public void close() throws IOException
    {
        throw new IOException("No experiment open");
    }
    public MDSDescriptor evaluate(String expression)
                                          throws IOException
    {
        throw new IOException("No experiment open");
    }
    public boolean isConnected()
    {
        return true;
    }
    public boolean isOpen()
    {
        return false;
    }
}
```

16.2.3 Managing State Transitions

The state transitions can be managed in the outer class. These are shown below:

```
//State Pattern Management. Each state is described by an
//inner class implementing the ServerState interface.

//State dependent methods. The behaviour of the following
//methods depends on the current connection state
public void connect(String source) throws IOException
{
    currState.connect(source);
    currState = connectedState;
}
public void disconnect() throws IOException
{
    currState.disconnect();
    currState = disconnectedState;
}
public void open(String experiment, int shot)
                                         throws IOException
{
    currState.open(experiment, shot);
    currState = openedState;
}
public void close() throws IOException
{
    currState.close();
    currState = connectedState;
}
public MDSDescriptor evaluate(String expression)
                                         throws IOException
{
    return currState.evaluate(expression);
}
// get and set and status methods
public boolean isConnected()
{
    return currState.isConnected();
}
public boolean isOpen()
{
    return currState.isOpen();
}
```

16.3 Our Final Product

The informal requirements for this revision might have read "convert the `DataServerFacade` class to be a state pattern". In the version of `EScope9` available on the CD, we have seized the opportunity to enhance the GUI to contain menu items for "Close" and "Disconnect". These options have always been available from the DataServerFacade and it is time to take advantage of them! We have also guided the user to select the correct menu item by greying out alternatives which are invalid in particular states. This is a very useful practice but it can also be a danger: the designer of the database connection protocol should not rely on a GUI designer getting it right!

Our state pattern implements the behavior shown in the state transition diagram shown in Fig. 16.2. As remarked earlier, there is a clear mapping between the diagrams such as this one and the state pattern. If new states are added to the diagram then new inner classes can be easily defined in the code. Similarly, new state transitions and new behaviors can be easily added to a state pattern. This is not the case for software which is coded with many conditional branches.

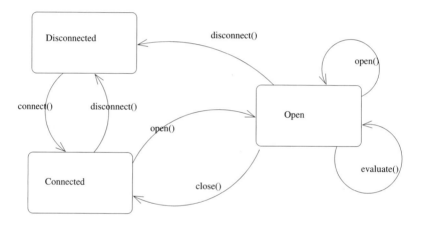

Fig. 16.2. Legal state transitions for the DataServerFacade object.

In general the state pattern has several advantages but it did end up being somewhat more complicated than the earlier version of DataServerFacade. Our advice is to save it for situations which have a complicated state transition behavior to begin with or which might be revised as requirements change. Grid-enabled eScience software which could be dependent on highly distributed computing and database resources might justify the use of a state pattern to negotiate and monitor resources.

17
Factory Patterns

Factories encapsulate constructors and they are normally used to separate the logic of construction from the target classes. We have already met factory methods several times in this book and the Java API is littered with them. Whenever you have factory methods you have some sort of factory pattern at work. In the Gang Of Four [28] book, an entire group of patterns is referred to as "creational" (or "factory" patterns). In the following section we will briefly describe features of some of these patterns.

17.1 A Factory Tour

Factory methods often start with the prefix "get" (sometimes "create"), and they can be confused with accessor methods. What distinguishes them is that the object being contacted will usually do something special to create the target object and it will usually not maintain an association with the target object.

17.1.1 Informal Factory Methods

If you find that a constructor, or a set of constructors, for a particular class is becoming too long-winded then it might be worthwhile replacing these constructors by calls to factory methods in another class. This is particularly so if the construction logic needs external parameters to be read and it contains branches based on the values of those parameters. The factory class can encapsulate all of the construction logic of the target class and make the target class less cluttered. These sorts of factory methods are related to the builder pattern.

In some cases it is quite natural for an object of one class to be responsible for the creation of an object of another class. As an example of this, you might have a data-structure and you might wish to create an object to step through that particular data-structure to access data elements. This would be an informal factory method but it generalizes to the "factory method pattern" of Java's `Iterator` hierarchy discussed below.

In Section 13.1.1 we observed that factory methods can be advertised in Java interfaces whereas constructors cannot. This gave rise to another informal use of factory methods. A factory method can reproduce a constructor inside a class in order to enforce a separation between software domains. A similar restriction in Java will be the motivation for introducing a factory method pattern into `EScope10` in this chapter.

Factory methods are sometimes *static*. A useful example of a static factory method will be demonstrated in Section 18.1.2 in the next chapter.

17.1.2 The Factory Method Pattern

The, so-called, "factory method pattern" is associated with two class hierarchies. The first one is a hierarchy of factory classes which contain factory methods to create objects from the other hierarchy of target classes. The factory hierarchy is usually descended from an abstract class or an interface. Depending on which concrete factory subclass is instantiated, different versions of the factory method will be made concrete and different objects from the target class hierarchy will be created. The pattern enables programs to be written that just use the interface of the target class. An object of a concrete target subclass gets constructed by the corresponding, concrete factory class. The decision of which concrete subclass to construct can be made at run time.

The `iterator` methods of the Java `Collection` framework are an example of a factory method pattern. Both the `Iterator` (target) hierarchy and the `Collection` (factory) hierarchy are descended from interfaces. The `Collection` interface defines a method, `iterator`, which should create an `Iterator` object to enumerate the elements of a collection. Concrete collections (lists, hash-tables etc.) create their own, corresponding, concrete iterator. You write your programs using just the `Iterator` interface and they will work regardless of which concrete iterator is instantiated. (Of course, there is much more to `Collection` classes than the `iterator` method. In this case the entire system of classes making up the collection hierarchy, the iterator hierarchy and the factory method is called an *iterator pattern*.)

17.1.3 Abstract Factory

The "abstract factory" method is poorly named. It would be better to call it a "family of factories" method. In this pattern, the client program associates with an interface which is the base level of a factory hierarchy. This interface will have several factory methods to create a family of target classes. The concrete factory class will then instantiate a particular family of products. The classic example of an abstract factory is a look-and-feel manager for GUI programming which can provide a factory method to return a whole family of factory methods to create each of the various widget classes needed for a particular GUI.

17.1.4 Builder

Some objects can have a very complicated representation. They may have many data attributes and some of these data attributes might be initialized in complicated ways.

The builder pattern separates the complicated construction of these objects from the target classes themselves. The builder class contains a set of methods which initialize the various attributes of the target class in the correct fashion. It then contains a factory method which will return the created object.

17.1.5 Prototype

The prototype pattern creates objects by cloning. That is basically all there is to it, except that you need to be careful of whether you wish to create a deep or a shallow clone. Shallow clones are quicker to create but object data instances are not copied. This means that you need to be careful of problems with shared references to one object.

17.1.6 Singleton

The purpose of the singleton pattern is to ensure that there is only one object of a particular class in a system. This can be important for resource management. It has an interesting structure with a private constructor. For example, we could use it if we wished to ensure that there was only one instance of the `GuiDomain` class in our `EScope` system. (There should only be one of these objects, but do we really care if someone creates another one??) The pattern might look like:

```
public class GuiFacade implements ....
{
    private static GuiFacade theGUI;
    ...
    private GuiFacade(){}
    public static synchronized GuiFacade getGuiFacade()
    {
        if (theGUI==null)
            theGUI = new GuiFacade();
        return theGUI;
    }
    ...
}
```

17.2 EScope10: Multiple Data Servers

In the next two revisions, we will extend the capabilities of `EScope` to handle multiple data servers. The scenario that we have in mind is that our browser might be used to download and explore data from a networked Grid of data servers. Furthermore, the database protocols and the nature of the data may be very different on different servers.

In the present revision, we imagine that a user would select a particular server name from a drop-down list. `EScope` would make the connection using "connect" and "open" dialogs as before. The amount of information solicited through dialogs would depend on how much was supplied in the `EScope.properties` file. This file could be edited and details of new servers could even be supplied to `EScope` *at run time*. Furthermore, a data server which requires a new protocol can be represented by a new data server package, and this package, in the form of `.class` files, could also be supplied to `EScope` at run time.

We will also use this revision to demonstrate that `EScope` can be used to visualize data from standard spread-sheets or simulation codes. In fact, if a given, specialized, data server domain has a facade class that implements the `DataServerFacadeInterface`, then there is no barrier to using `EScope` to interactively explore all sorts of data – not just the fusion waveform data considered here.

In this revision, to `EScope10`, a factory method pattern will be employed to create `dataServerFacade` objects as they are requested by a user.

17.2.1 Requirements

This revision is more on the "fictitious" side than our earlier examples. Our informal requirements for it might read

1. Users can specify the following details of data servers in the properties file:
    - A name for the server to appear in a selection list.
    - An "address" for the `.class` file which will handle interaction with a particular server.
    - Optionally, an internet address and port number for the server.
    - Optionally, the experiment and shot number for a set of data to be displayed.
    - Any parameters which need to be passed to the data server on start up.
2. When the "connect" menu item is selected, the user will be shown a list of available servers. When a server is selected, then one of two things will happen:
    - If the internet address and port number have been supplied in the properties file then a connection will be made. Additional parameters in the properties file, if any, will be passed to the data server.
    - If no internet address is specified, then the connect dialog will be shown and a connection will be made when it is complete. Additional parameters in the properties file, if any, will be passed to the data server.

3. Following connection, the open dialog will be shown. Default values for the experiment and shot number will be displayed in this dialog.

In addition to these requirements, we have chosen to demonstrate the capability of `EScope` to interface with a quite different data server from `MDSplus`. A screen shot of `EScope10` visualizing hypothetical spread-sheet data is shown in Fig. 17.1 and the listing of the `TextDataServer` class, which connects to this data, can be found in Appendix F.5.4 on page 370. (`TextDataServer` is discussed below in Section 17.2.6.)

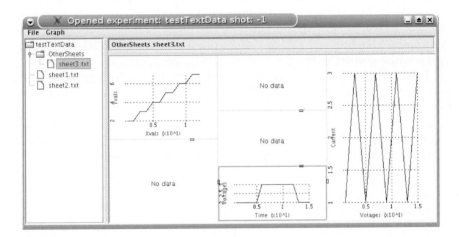

Fig. 17.1. A screen shot of `EScope10` showing multiple waveforms from a file of readable text data. The data are arranged in columns of X and Y values and were output from a standard spread-sheet application.

17.2.2 Implementation

Many of the above requirements are concerned with the GUI domain. These include the addition of a new dialog class to be able to choose the server to connect to. The logic of the `GuiFacade` will need to be modified to meet the new requirements and an expanded protocol for the properties file will need to be specified. Outside of the GUI domain, a new `TextDataServer` domain will also need to be supplied.

Rather than justifying the way in which this refactoring step might be accomplished, we plan to jump straight into describing our solution code, `EScope10`, which is available on the CD. It turns out that all of the changes can be accomplished by adding additional classes and the modification to the `GuiFacade` is mostly localized to the `ConnectAction` inner class.

17.2.3 Example Properties File

An example properties file for use with `EScope10` is shown below. The first group of entries specify the configuration of the graph windows (as discussed in Section 13.1.1). This is followed by a number of entries for each data server. The data server groups are numbered and must include a name for the server and a relative address for a class which will make the connection to that server (in our example these are factory classes). The remaining property items, "Source", "Experiment", "Shot" and "Args" are optional. They can be specified in order to avoid repeatedly typing them into the connect and open dialogs. Specifying `Source` causes the connect dialog not to be shown, so it is worthwhile to have a blank entry for this property if a network connection does not need to be made. (We have done this for the `TextDataServerFactory` entry in the properties file.)

It is worth noting that a properties file might be available from a system manager so ordinary users would not need to know about the network address, port number, or the factory classes. The file might even be written by a computer program! This could be the case in a dynamic, Grid-based application where servers could announce their availability and provide information, and even the software, to enable connections to be made to them.

```
num_columns=3
column_1.num_rows = 2
column_2.num_rows = 3
column_3.num_rows = 1

#DataServer names and factory class names must be specified
#Internet address and port number are optional
#Experiment and shot are optional
#Factory constructor arguments are optional
DataServer_1.Name = MDSplus
DataServer_1.FactoryClass=
                    mdsServerDomain.MDSDataServerFactory
DataServer_1.Source = ephebe:8000
DataServer_1.Experiment= h1data
DataServer_1.Shot = 37025
#DataServer_1.Args = 100000

DataServer_2.Name = Text
DataServer_2.FactoryClass=
                    textServerDomain.TextDataServerFactory
DataServer_2.Source=
DataServer_2.Shot =
DataServer_2.Experiment = testTextData
```

17.2.4 The `ServerSelectDialog`

A listing for the new `ServerSelectDialog` class is given in Appendix F.5.1 on page 364. It is notable for the way in which the properties file is read and processed. Note that we have chosen to load and read this file in two places in the overall program: initially to set up the graphics windows and every time that a `ServerSelectDialog` is created. In the `ConnectAction` inner class of the `GuiFacade`, shown in Appendix F.5.2, on page 368, it can be seen that we create this object each time the connect menu item is chosen. This means that the server configurations can be edited while EScope10 is running. In fact, new data server domains, represented by compiled Java packages with the correct relative package address, can be attached at run time. (These class files could even be downloaded, dynamically, from a remote server.)

17.2.5 The Factory Pattern in EScope10

In Java, it is possible to instantiate an object at run time starting from a `String` containing the class name for that object. Every Java class has an instance of another class called `Class` which is created by the Virtual Machine and which represents the class while a program is running. The `Class` class has a static method, `forName`, which will take a class name (with package address) as an argument and return the actual `Class` object which represents it. Once this `Class` object has been obtained, another, non-static method called `newInstance` can be called to create an object of the named class itself! (This sounds complicated, but it becomes easier to understand after following an example such as the code excerpt shown below.)

Invoking the `newInstance` method is equivalent to calling a constructor for a given class without arguments. This would be satisfactory if we could be assured that any data server constructor we wished to use would have no arguments. In order to get around this restriction, we employ the factory method pattern: `ConnectAction` creates a factory class. This factory class will have a constructor with no arguments. It will have a `factory method` which will accept a generic argument list (which we chose to be an array of strings) and which will return an object which implements `DataFacadeInterface`.

The relevant excerpt from `ConnectAction` is shown below. The full listing is given in Appendix F.5.2 on page 368.

```
String factoryClassName = serverD.getFactoryClassName();
try
{
    DataServerFactoryInterface currFactory =
            (DataServerFactoryInterface)
            (Class.forName(factoryClassName).newInstance());
    String[] args = serverD.getArgs();
    database = currFactory.createDataServer(args);
}
```

The factory interface and factory methods are listed in Appendix F.5.3, on page 369, and a class diagram, ignoring domains, is shown in Fig. 17.2. When dealing with the full package structure of `EScope`, we choose to change the name of our old `dataServerDomain` to `mdsServerDomain`. (To do this we simply change the directory name and modify the package statements in the header of each file. There are no other changes to make in the rest of the software!)

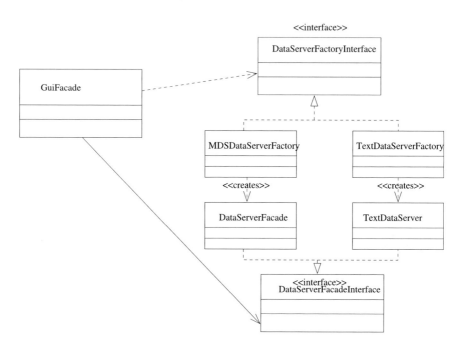

Fig. 17.2. The implementation of the factory pattern between the GUI and the two data domains is shown ignoring packages.

17.2.6 A Text Data Server

The idea of our text data server is to provide an interface to a directory tree containing text files with data. In each text file, data is assumed to be defined in two columns for the X and Y values. A title for each column can also be defined. We provide a simple example of such a directory tree in the file `testTextData` which is include with `EScope10` on the CD. These sorts of data files can be generated by mainstream, spread-shcet applications as well as traditional simulation codes.

The implementation of the class `TextDataServer`, shown in Appendix F.5.4, on page 370, is not particularly complicated. It implements `DataServerFacadeInterface` as follows:

- The methods `connect`, `disconnect` and `close` are empty.
- The method `open` uses only the "experiment" argument, which is interpreted as the path name of a top directory. The directory structure is then traversed, collecting file information in a tree structure formed of `javax.swing.tree.DefaultMutableTreeNode` objects. This tree structure is then embedded in a `javax.swing.tree.DefaultTreeModel` object. The method `constructTree()` causes the files and directories to be displayed in the EScope GUI.
- The method `getPlotData()` receives as argument the the path name of the corresponding text file and reads the X and Y values, and possibly the associated titles, which are then returned in a `GraphData` object. The `GraphData` class in the `textServerDomain` is identical to that in the `dataServerDomain`.

17.3 Our Final Product

This revision has demonstrated some important principles:

1. `EScope` can be readily, and usefully, generalized to data servers with a different protocol to `MDSplus`. Our example read a text file; a useful extension would be to adapt the code to process XML tags to define the data arrays.
2. The factory method pattern can be used together with property files to let `EScope` switch and select between data servers.
3. The architecture of `EScope` enables us to "bolt on" a new data server domain without affecting the rest of the program. This bolting-on could even be done at run time.

In spite of its increase in functionality, the changes needed to produce `EScope10` have been mostly limited to some simple factory classes and an inner class of the GUI facade.

Although this revision demonstrates some important principles, it is, admittedly, artificial because our justification for using the factory method pattern was to pass arguments to the constructor of the data server facade but the parameter which we passed was not really used. What we did was to pass a parameter to the `mdsServerDomain.ProxyDataServer` factory method which, ostensibly, could have been used for specifying the maximum number of datasets to store in the proxy cache. A full implementation of this option, or of another caching policy (see Section 15.1), could be a useful exercise for the reader.

18

Chain of Responsibility

Like the decorator pattern, the chain of responsibility pattern uses a chain of classes. In this case one or more members of the chain is potentially able to handle a request from a client. When a request is made, it travels down the chain until the first handler object is found.

18.1 `EScope11`: Avoiding Explicit Connection to Data Servers

The scenario we have in mind for this revision is slightly more dynamic than for `EScope10`. Imagine that `EScope` is a portal onto a Grid of potential data servers. Rather than requesting a particular data server to be searched for data, a request could be made to the complete ensemble of servers. The traditional chain of responsibility pattern would choose the first eligible data server to establish the connection. Depending on what policy you wished to subscribe to, you might wish to vary this to search the entire ensemble until a particular criterion (such as download time) had been optimized.

18.1.1 An Example Properties File

The following excerpt is typical of a properties file for this revision. Compared with `EScope10`, the servers are no longer named because there will not be a server selection dialog. There is also only one entry for the default experiment and shot:

```
#Factory class names, and internet address
#and port number are mandatory
#Factory constructor arguments are optional
DataServer_1.FactoryClass=
                    mdsServerDomain.MDSDataServerFactory
DataServer_1.Source = localhost:8002
```

```
DataServer_1.Args = 100000
DataServer_2.FactoryClass=
                        textServerDomain.TextDataServerFactory
DataServer_2.Source=
#Default experiment and shot for Open dialog
DataServer.Experiment= h1data
DataServer.Shot = 37025
#DataServer_2.Shot =
#DataServer_2.Experiment = testTextData
```

18.1.2 Implementation

Compared with EScope10, the refactoring to EScope11 is about *taking classes out*! There will be no server select dialog and no connection dialog. The Connect menu item will be replaced by an OpenConnect item which will create the server chain and then show the open dialog with any defaults. (The Open menu would not create the server chain.)

The main addition is a new class, DataServerHandler whose listing is given in Appendix F.6.1 on page 375. There are several points to note about this particular implementation of the chain of responsibility pattern:

- This class uses a *static factory method* to create the list of data handlers. This means that the constructor can be used to create one member of the chain and the factory method can be used to create the entire chain.
- Each data server is connected to as it is added to the chain. This policy could be time-consuming in a real Grid application with remote servers.
- A connection is made when a particular experiment is opened. If an exception is thrown then the next data server on the list is used. This policy could be varied to, for example, check a particular information file to see whether an experiment dataset should be found on a particular server before opening it. Alternatively, a combination of policies could be used where some servers publish information sheets and others are tested by opening them.

In order to maintain a strict separation between interfaces and implementation, the DataServerHandlerInterface, which defines the method openExperiment(..), has been placed in the sharedInterfaces package. This hides the actual implementation of DataServerHandler from the client classes. Using this approach, the changes required to package guiDomain are minimal and mostly represented by the *removal* of code.

18.1.3 Our Final Product

This revision can be thought of as a variant of EScope10. Together they provide a starting point to think about how EScope might be effectively ported to the dynamic e-Science environment of the Grid.

19
Design Patterns and Threads

Although we have briefly considered threads in Section 2.5, we have yet to use them in `EScope`. This might suggest that thread programming is not so important in real-world e-Science applications. In fact, nothing could be further from the truth, especially when networked applications using the Grid are concerned.

From the discussion of Section 2.5, it is inconceivable that a Java TCP/IP server would not employ multi-threading. There are other common situations in which multi-threading can be used to improve the quality of a user-interface. For example, graphical user interfaces respond to events such as window resizing, key presses, and mouse motion. Event management is typically carried out by a single thread that repeatedly reads from an event queue. If a lengthy computation is triggered by an event, an unthreaded application will not react to other events until this computation has terminated and this will result in the application becoming blocked. Graphical interfaces which block can make users very nervous!

Some readers will have already experienced this blocking behavior when they tried to run `EScope` and selected large signals for visualization. Even if `EScope` might not be able to perform very useful tasks at the same time that a large signal file was being downloaded, giving it a way of reacting to other mouse clicks could greatly improve the perceived quality of its user interface. In addition, providing a progress bar which indicates the current download status would make the user confident that something useful was happening. In the present revision of `EScope`, we will implement such a progress bar.

19.1 Threads and Race Conditions

When dealing with multiple, threaded, flows of execution, sequential program logic is not guaranteed and it is necessary to predict the possible *race conditions* which might occur.

Just for a taste, we will consider a standard example of a race condition. Consider a very simple concurrent program which defines two threads. Both threads share a buffer object which can store one integer. The buffer also contains a flag, `empty`, which is set to `false` whenever an integer is stored. If an integer is read from the buffer, then `empty` is set to `true`. The first, "producer" thread checks whether the buffer is empty. If it is empty, then the producer writes an integer value to the buffer which causes `empty` to be set to `false`. The second, "consumer", thread waits for a new datum and repeatedly checks the value of `empty`. When `empty` is `false` it reads the datum which causes `empty` to be set to `true` once again. A Java implementation for this is provided in the Listing 19.1. The classes `Producer` and `Consumer` inherit from `java.lang.Thread`, and their `run()` methods represent the two independent flows of execution in the program.

Listing 19.1. Example of a race condition using a shared data buffer.

```
public class TestProducerConsumer
{
    public static void main(String args[])
    {
        Buffer buffer = new Buffer();  //The shared buffer
        Producer producer = new Producer(buffer);
        Consumer consumer = new Consumer(buffer);
        //Start threads
        producer.start();
        consumer.start();
    }
}
class Buffer //The shared data buffer
{
    private int data = 0;
    private boolean empty = false;

    public boolean isEmpty() {return empty;}
    public int getData()
    {
        empty = true;
        return data;
    }
    public void putData(int data)
    {
        empty = false;
        this.data = data;
    }
}
class Producer extends Thread
{
    private Buffer buffer;
    public Producer(Buffer buffer)
```

```java
    {
        this.buffer = buffer;
    }
    public void run()
    {
        int count = 1;
        for(int i = 0; i < 10; i++)
        {
            while(!buffer.isEmpty()) {}
                    //Wait until data has been read
            buffer.putData(count);
            count++;
        }
    }
}
class Consumer extends Thread
{
    private Buffer buffer;
    public Consumer(Buffer buffer)
    {
        this.buffer = buffer;
    }

    public void run()
    {
        int readData = -1;
        for(int i = 0; i < 10; i++)
        {
            while(buffer.isEmpty()){}
                    //Wait until data has been written
            readData = buffer.getData();
            System.out.println(readData);
        }
    }
}
```

When the program in Listing 19.1 runs (it is provided on the accompanying CD), we would expect that the producer thread writes increasing values in the shared buffer and the consumer thread collects these values and prints them. The result should be that the numbers from 0 to 9 get printed onto the screen. It is very possible that you actually do observe this correct behavior but you will be surprised at how slow it is despite the very limited number of instructions needed to print 10 numbers on a computer screen. The reason is that most of the time the program loops while checking the `empty` flag. If, for example, the producer is looping while waiting for the `empty` flag to be set to `true`, then the consumer will not have a chance to read the number until the operating system passes the control of the CPU to it. This means that a lot of CPU cycles are wasted by performing useless operations. Even worse, we cannot be absolutely sure that the program will print out the correct sequence of numbers! For example, it may happen that the producer thread could be interrupted just after setting the `empty` to be `false` and before writing a new number to the buffer. The consumer thread might then re-read an old datum.

This is a typical race condition. It is an execution sequence which was probably not considered when writing the program, but which may occur "randomly" during execution. There are counter-measures to race conditions, and we will discuss them after first describing two tools that Java provides us for handling concurrency in programs: synchronized methods and `wait()`/`notify()`.

19.2 Synchronized Methods and `wait()`/`notify()`

Synchronized methods are declared using the keyword `synchronized`. The Java virtual machine ensures that an object with synchronized methods is locked whenever one of its synchronized methods is called. If another thread tries to call *any* method for the same object, it is suspended, queued and put into a *wait state* until the owner thread exits the synchronized method. Once the owner thread has finished with the synchronized method, then one of the threads which has been queued for access to the object is selected based on the priority associated with each queued thread. The way in which the Java virtual machine chooses between threads of the same priority can be system dependent and, in the past, threaded Java applications have not been the best demonstration of portability! (Starting from Java 1.5 there is a new concurrency library in `java.util.concurrency` which should improve the performance and portability of threaded applications.)

The methods `wait()` and `notify()` are defined for the `Object` class and are available for all classes. Both methods require that the current thread is the owner of the object, i.e. they have to be called from within synchronized methods. The `wait()` method suspends the current thread *and releases the ownership of the object*. The current thread is then put into the queue of waiting threads for that object. The `notify()` method awakes one of the waiting threads and that thread resumes execution on the shared object from the programming line immediately following the `wait()` call which suspended it. The `notifyAll()` method awakes *all* the threads waiting for an object and these threads now compete for access to that object based on their priority. (You should use `notifyAll()` unless you are sure that all queued threads are waiting for the same condition.)

To convince ourselves that `wait()` and `notify()` do what we need, we shall use them to improve our producer-consumer program. First of all, we observe that there is no need to keep the producer thread busy checking the value of `empty` once it has already been read as being `false`. Instead, it suffices to suspend execution of the producer thread until the buffer really is empty. Similarly, the consumer thread can be suspended once `empty` has been read as being `true`. If we do this, the suspended threads will need to be awakened

19.2 Synchronized Methods and wait()/notify()

when the condition for which they are waiting is satisfied. This is achieved by calling `notify()` just after writing (by the producer) or reading (by the consumer).

The correct code is shown in Listing 19.2. In this version, all of the waiting and notifying is done inside the synchronized `Buffer` methods `getData()` and `putData()`. This ensures that the datum is returned by `getData()` only when it is available. Otherwise the calling (consumer) thread is suspended until `notify()` is called from within `putData()`. The same holds when the producer task is suspended because the datum has not been read. In this case it will be resumed by the `notify()` call from within `getData()`. Note that a race condition cannot occur here because both methods `getData()` and `putData()` are declared as `synchronized` so that the thread checking the `empty` flag cannot be interrupted before reading or writing the datum.

Listing 19.2. A shared data buffer using synchronized methods.

```
class Buffer1  //The shared data buffer
{
    private int data = 0;
    private boolean empty = false;
    public synchronized int getData()
    {
        while(empty)
        {
            try
            {
                wait();
            }catch(InterruptedException exc){}
            //wait() may generate an exception when the
            //thread is interrupted while waiting
        }
        empty = true;
        notify();
        return data;
    }
    public synchronized void putData(int data)
    {
        while(!empty)
        {
            try
            {
                wait();
            }catch(InterruptedException exc){}
        }
        empty = false;
        this.data = data;
        notify();
    }
```

```
}
class Producer1 extends Thread
{
    private Buffer1 buffer;
    public Producer1(Buffer1 buffer)
    {
        this.buffer = buffer;
    }
    public void run()
    {
        int count = 1;
        for(int i = 0; i < 10; i++)
        {
            buffer.putData(count);
            count++;
        }
    }
}
class Consumer1 extends Thread
{
    private Buffer1 buffer;
    public Consumer1(Buffer1 buffer)
    {
        this.buffer = buffer;
    }
    public void run()
    {
        int readData = -1;
        for(int i = 0; i < 10; i++)
        {
            readData = buffer.getData();
            System.out.println(readData);
        }
    }
}
```

19.3 Patterns for Concurrent Systems

The above discussion may have convinced the reader that programming with threads is very complicated. Fortunately, there are patterns which can help us and we have just described one in the previous example. Design patterns for threads were not included in the seminal GoF book [28], but there has been much subsequent work in this area (see, for example, [35, 36]).

19.3.1 The Acceptor-Connector Pattern

The first concurrent pattern example that is worthy of mention is the threaded server which we discussed in Section 2.5 and whose listing is shown in Section 2.5.1. This example is a version of the *Acceptor-Connector* pattern. On the server side there are two classes. One class establishes the connection with the clients and runs in the main server thread. The instance of this connection class provides only the *establishment* of the service, and it defers the specific service *provision* to a newly created instance of the second class. Instances of this second, "handler", class receive, in their constructors, the required information about the connection and they carry out the service for the client in a separate thread. The entire server program has $n + 1$ threads when handling n clients. The pattern decouples the connections between services over a network from the processing carried out once connections have been made.

19.3.2 The Asynchronous Method Pattern

Another common pattern in concurrent systems is one which we will call the *asynchronous method* pattern. It will be used in the refactoring to produce EScope12 which will have a progress bar for retrieving datasets from the data server.

In our discussions so far, we have been talking about the concurrency of methods associated with particular class instances (objects). An instance of a given class is able to provide a set of services, and each of these services is represented by a method. Conventionally, when a method is called the caller thread waits for the termination of the method before executing the next statement. But in the asynchronous method pattern the service is provided by a thread which is different from the caller thread and the caller can resume execution before the method has been completed. The two threads, caller and called, can execute in parallel.

In order to handle the termination of an asynchronous method, a number of cases need to be considered:

1. The caller thread could continue indefinitely in parallel with the called thread if the actions performed by the called method are fully independent. In this case no notification mechanism is required.
2. The caller thread could continue in parallel with the called thread until the called thread terminates. But after the called thread terminates, some specific actions are required by the caller thread.
3. The caller thread could perform a certain number (possibly zero) of actions in parallel with the called thread, but eventually it needs to suspend its execution until the called thread terminates.

One mechanism to handle the termination of an asynchronous method is to pass a *notifier* object to the called thread. Both caller and called threads maintain associations with this object and it can be used for communication between them. In a common implementation of this pattern, a notifier class would define, at least, the following methods:

- `executionTerminated()` would be invoked by the called thread when it is about to terminate.
- `waitForTermination()` would be invoked by the caller thread to suspend its execution until the called thread terminates.

A possible implementation of such a notifier class is the following:

Listing 19.3. AbstractNotifier class.

```
abstract class AbstractNotifier
{
    private boolean terminated = false;
    public abstract void specificExecutionTerminated();
    public synchronized void executionTerminated()
    {
        specificExecutionTerminated();
        terminated = true;
        notifyAll();
    }
    public synchronized void waitForTermination()
    {
        while(!terminated)
        {
            try
            {
                wait();
            }catch(InterruptedException exc) {}
        }
    }
}
```

In Listing 19.3, the `AbstractNotifier` class carries out the generic functionality required for notification and synchronization. Concrete subclasses will implement the `specificExecutionTerminated()` method.

The `AbstractNotifier` class mediates communication between calling and called threads in the following way. When the calling thread invokes `waitForTermination()`, sometime after calling the asynchronous method, either one of the following conditions holds:

1. The called method thread has already finished execution. In this case the flag `terminated` is `true`, and `waitTermination()` returns immediately.
2. The called method thread has not yet finished execution, and the flag `terminated` is `false`. In this case, the caller thread calls `wait()` and suspends execution until the called thread calls `notifyAll()` within `executionTerminated()`.

Readers may be interested to know that the design of the `Abstract-Notifier` class can be *proven to never give rise to a race condition*. If the two methods in this class were not `synchronized`, then it would be possible for the caller thread to have tested the value of `terminated` and found it to be `false` and be about to call `wait()`. In the meantime the called thread could have set `terminated` to `true` which would mean that the caller would suspend indefinitely. But making the methods `synchronized` removes this possibility. The full proof is slightly technical and we omit the details here. In practice the development of proofs for every concurrent program would not be feasible so there is a clear advantage in following a design pattern which has already been proved to be correct!

19.3.3 More Complete Implementations of `AbstractNotifier`

Complete implementations of `AbstractNotifier`, also need to handle the case in which the caller is no longer interested in the execution of the asynchronous method. This happens, for example, if it is assumed that the request may have failed after a certain amount of time. A notification of termination which arrives after this amount of time may then produce unwanted effects.

The more complete implementation of `AbstractNotifier`, in Listing 19.4, includes an `invalidate()` method which, possibly, awakes waiting threads after a timeout. When this happens, `waitForTermination()` will return `false`, indicating that the request has been invalidated.

Listing 19.4. Abstract notifier class.

```
abstract class AbstractNotifier
{
    private boolean terminated = false;
    private boolean valid = true;

    public abstract void specificExecutionTerminated();
    public synchronized void executionTerminated()
    {
        specificExecutionTerminated();
        terminated = true;
        notifyAll();
    }
    public synchronized boolean waitForTermination()
    {
        while (!terminated && valid)
        {
            try
            {
                wait();
            } catch (InterruptedException exc) {}
        }
        return valid;
```

```
    }
    public public synchronized void invalidate()
    {
        valid = false;
        notifyAll();
    }
}
```

Timer functionality can be incorporated into `waitForTermination()` using the `Timer` class from the `java.util` package. `Timer` contains a method `schedule()` which takes two arguments: an implementation of the interface `TimerTask` and a delay time (in milliseconds). `TimerTask` defines a single method `run()` which is called when the specified timer interval has expired. In the following listing, `run()` calls `invalidate()` which, in turn, awakes possible threads waiting for termination. Note that `run()` is executed in yet another thread which is activated by the timer.

Listing 19.5. More sophisicated `waitForTermination` method.

```
synchronized boolean waitForTermination(long timeout)
{
    Timer timer = new Timer();
    timer.schedule(new TimerTask ()
        {
            public void run()
            {
                invalidate();
        }}, timeout); //end of call to schedule

    timer.start();

    while(!terminated && valid)
    {
        try
        {
            wait();
        }catch(InterruptedException exc) {}
    }
    return valid;
}
```

Concrete subclasses for `AbstractNotifier` are normally implemented as inner classes of the caller class, so that the `specificExecutionTerminated()` method can access internal fields. `AbstractNotifier` subclasses may also add other methods to monitor the current status of the asynchronous method execution. We shall see such an example in `EScope12`, where a status bar will monitor the current status of downloading a signal file from the data server.

19.3.4 Other Classes in the Asynchronous Method Pattern

A possible implementation of the other classes defined in the asynchronous method pattern is provided in Listing 19.6. The class `AsynchExecutor` defines a method `spawn(..)` which has, as arguments:

1. an object which contains the method to be executed,
2. a set of arguments to that execution method, and
3. a notifier object.

The method to be executed is passed as an implementation of the `Executor` interface. This interface defines a single method `execute(..)`. The arguments for `execute(..)` are also passed through `spawn(..)` as a single `Object` instance. `execute(..)` is called in the `run()` method of the `ExcecutorThread` instance created by method `spawn()`. Note that it is necessary to pass arguments to the constructor of `ExecutorThread` because it is not possible to pass them directly to `run()`.

Listing 19.6. Other classes from the asynchronous method pattern.

```
interface Executor
{
    public void execute(Object arguments,
                        AbstractNotifier notifier);
}
class AsynchExecuter
{
    public static void spawn(Executor executor,
            Object arguments, AbstractNotifier notifier)
    {
        ExcecutorThread execThread =
            new ExecutorThread(executor, arguments,
                               notifier);
        execThread.start();
    }
}
class ExecutorThread extends Thread
{
    private Executor executor;
    private Object arguments;
    private AbstractNotifier notifier;
    public ExecutorThread(Excecutor executor,
            Object arguments, AbstractNotifier notifier)
    {
        this.executor = executor;
        this.arguments = arguments;
        this.notifier = notifier;
    }
    public void run()
    {
```

```
            executor.execute(arguments);
            notifier.executionTerminated();
        }
    }
}
```

19.3.5 Summary of the Asynchronous Method Pattern

In summary, the steps defined in the Asynchronous Method design pattern are the following:

1. The caller creates an implementation of the `Executor` interface by wrapping up the desired asynchronous method call inside the `execute()` method defined by this interface.
2. The caller creates an object which implements `AbstractNotifier` and which provides an implementation of the `specificExecutionTerminated()` method to handle the termination of the asynchronous method.
3. The caller calls the `AsynchExcecutor.spawn()` static method and passes the `Executor` object, the argument object and the notifier object to it.
4. `AsynchExecutor.spawn()` creates an instance of `ExecutorThread` and starts the thread. This thread's `run()` method calls the `Executor`'s `execute()` method and passes the argument object and the notifier object to it. (Note that the notifier object may be used by the `execute()` method to, for example, report the current status of the execution. In this case, the concrete `AbstractNotifier` subclass would define additional reporting methods.)
5. When `execute()` terminates, the notifier's `executionTerminated()` method is called.

19.3.6 The Active Object Pattern

In the asynchronous method pattern a new thread is created whenever the asynchronous method is called. A similar pattern, known as *active object*, differs from asynchronous method in that the same thread is used for subsequent method calls. Active object is the preferred pattern if the operations carried out by the asynchronous method cannot occur in parallel with other flows of execution for the same method. It is more complicated than the asynchronous method pattern because there is a need for some kind of queueing mechanism for handling calls to the asynchronous method which get made before a previous invocation has terminated.

In the next section, we shall see that, in `EScope12`, it is possible for the asynchronous method to be called again before a previous invocation has terminated. However, because this application uses the asynchronous method for downloading files, and its invocation is activated in response to mouse clicks on nodes of the file tree, the number of pending threads is likely to be limited. Therefore, the simpler asynchronous method pattern is used instead of active object.

19.4 EScope12: A Progress Bar for Downloading Signals

As discussed above, a GUI such as EScope can become frozen during a time-consuming operation such as downloading a large signal file from the server. The result is that the interface becomes less "fluid" to use – thus worsening its quality as well as worsening the mood of the user!

Even if download times cannot be reduced by clever programming, it is possible to provide an acceptable compromise in the form of a progress bar. If a user can see a progress bar then she or he will be confident that the application is doing something useful. The quality of the interface will be even further increased if it will still be able to respond to mouse events and, better again, allow the manipulation of the already-downloaded waveforms.

To provide a progress bar we need to introduce multi-threading into EScope, and the asynchronous method pattern will be used to decouple the downloading of signals from the rest of the system. Before introducing threads, let us reorganize GuiFacade so that the code of the call-back method valueChanged(), in the inner class DataTreeSelectionListener, gets embedded in a separate method called executeValueChanged(). This method will receive the original TreeSelectionEvent argument, plus a reference to an NotifierInterface implementation. The NotifierInterface argument will then be passed to the getPlotData() method of DataServerFacade and thence to the constructor of MDSMessage. Observe that, up to now, we have only made some mechanical rearrangements of the code with no effect on the behavior of the system. NotifierInterface is defined as follows:

```
public interface NotifierInterface {
    public boolean waitForTermination();
    public void executionTerminated();
    public void reportProgress(float percent);
    public void reportMessage(String message);
}
```

The first two methods in NotifierInterface are those implemented in AbstractNotifier as described above. The last two methods are used by MDSMessage to communicate the current status of the downloading operation to the concrete notifier object, PlotDataNotifier, which is an inner class of GuiFacade). This information is communicated by updating a JProgressBar and a JLabel. Once the code has been reorganized the adoption of the rest of the asynchronous method pattern is straight-forward.

Because the new classes required for the asynchronous method pattern will be used only within the guiDomain, we place them, and three of the interfaces, into that package. NotifierInterface is put into the sharedInterfaces package because it is also used by classes in the mdsServerDomain.

19.4.1 Using Threads with Swing

There is a further subtlety which we need to consider before our implementation is complete. The last action of method `executeValueChanged()` can be represented by the following three lines:

```
GuiFacade.this.setDescription(shortFileAddresses);
GuiFacade.this.graph.setGraphData(graphs);
GuiFacade.this.graph.repaint();
```

These calls are performed by a separate thread and involve the activation of Swing methods. Most methods in Swing are, however, *not thread safe*. That is, they behave correctly only when called by the main Swing thread. The method `setGraphData()` invokes many graphical operations since it has to plot the passed signal. The results are, in this case, unpredictable and it is very likely that, after a few draw operations, crazy things will happen! Swing provides a solution to this problem via the `SwingUtilities.invokeLater()` static method. This method enables a portion of code, embedded in a `Runnable` object, which is passed as an argument, to be executed by the main Swing thread. For this purpose, the `executeValueChanged()` method passes an instance of the class `GraphPainter`:

```
SwingUtilities.invokeLater(
        new GraphPainter(graphs, shortFileAddresses));
```

The `GuiFacade` inner class `GraphPainter` is shown below:

```
class GraphPainter implements Runnable
{
    private GraphDataInterface[] graphs;
    private String shortFileAddresses;
    public GraphPainter(GraphDataInterface[] graphs,
                        String shortFileAddresses)
    {
        this.graphs = graphs;
        this.shortFileAddresses = shortFileAddresses;
    }
    public void run()
    {
        GuiFacade.this.setDescription(shortFileAddresses);
        GuiFacade.this.graph.setGraphData(graphs);
        GuiFacade.this.graph.repaint();
    }
```

As a final remark, we note that the use of the asynchronous method pattern in `EScope12` allows the manipulation of other waveforms while downloading a new waveform. You can try this yourself using the version of this code available on the CD. If you run it using the `MdsipSimulator` server (which is useful for this example because downloads will be much slower than for the `MDSplus` server) you should be able to grab, zoom and apply the cross-hair diagnostic as the selected dataset is downloading. A further improvement to the software, which is left as an exercise to the reader, would be to enable two datasets to be downloaded at the same time.

19.5 Programming Exercises

Try running the `TestProducerConsumer1` program supplied on the CD and verify that it is much faster than `TestProducerConsumer`. Experiment with it to get the idea of how synchronized methods and `wait()`/`notify()` work together. For example, comment out the calls to `notify()` and verify that the program does not work properly.

19.6 Further Reading

It should be apparent that the discipline of writing good, multi-threaded software is a very demanding one. As discussed in the next chapter, server-side e-Science applications will typically require support for distributed and heterogeneous computing and data platforms such as the Grid. For these systems, multi-threaded software is of interest and the reader would do well to consult authoritative discussions of patterns in concurrent systems in the literature, on the Web, and in books such as [35, 36].

20
Postscript

20.1 Design Patterns Then and Now

In 2006, the *Association Internationale pour les Technologies Objets* awarded the Dahl-Nygaard Prize to Erich Gamma, Richard Helm, Ralph Johnson and John Vlissides[1], collectively known as the "Gang of Four". This award was presented at the ECOOP 2006 conference in Nantes, France in July 2006. In the opening keynote presentation to this conference, Erich Gamma spoke on the topic of "Design Patterns - 15 Years Later". He made several points in this talk which are very relevant to the `EScope` case study described in this book[2]

1. Design patterns are best thought of as targets for refactoring.
2. Not all of the classic 23 design patterns from [28] are equally useful. In fact, not all of them would be included as "core patterns" in a revision of the book.
3. It is difficult to learn design patterns from a book, but having a book is better than not having a book!
4. What is really needed are books which describe the evolution of a real world case study as it is built up and then refactored using judiciously-chosen design patterns.

This last of Gamma's points is exactly what we have attempted to achieve in this book. Our personal experience is that design patterns really do work in this manner. The patterns that we have illustrated here have enabled us to build a software architecture for `EScope` which has been reasonably robust and extensible.

Although we have been happy enough with `EScope` to recommend it as a case study for others, we note that there is quite some scope for interpretation of which patterns should be used at which stages in a refactoring process and

---

[1] Sadly, John Vlissides passed away in November 2005.
[2] These points have been quoted with permission from Erich Gamma.

how these patterns might best be implemented. Others may agree or disagree with pattern choices and implementations that we have described. In fact, there is a human cost to spending time implementing design patterns and readers may, at the end of the day, decide not to follow all of the ideas presented here. The advantages and disadvantages of adopting particular design patterns are described in more detail in [28] and other references.

20.2 The e-Science "Software Stack"

Like any area of rapid technological advance, e-Science is full of buzz-words and acronyms. Although a full treatment of the contemporary concerns of e-Science is outside the scope of this book, we shall attempt to give something of an overview by describing the dominant themes from the First International Conference on e-Science and Grid Computing which was held in Melbourne, Australia, in December 2005 [37]. For a brief introduction to any of the buzzwords below, we recommend the Wikipedia (http://en.wikipedia.org/). (Parenthetically, many of the papers at this conference described software systems which were built using Java technology and it is clear that a study of Java is a good route to gaining programming expertise for e-Science!)

The dominant themes at this conference were

- **Applications**
  The applications described included the development of Grid portals for scientific software. Sometimes this was achieved using environments such as GridSphere (www.gridsphere.org) which is an open-source portlet-based framework. Portlets are Java-based web components. Applications also included volunteer and community-based Grid computing, the development of collaborative frameworks and the integration of data stores on the Grid. Specific papers described:
  - A volunteer computing framework for climate prediction.
  - An XML schema mechanism for integrating DNA sequence databases on a network together with computing resources and query services.
  - Grid-enabling of legacy applications using toolkits. The wrapping of legacy applications as Web Services.
  - Collaborative frameworks for digital object repositories.
  - A chemistry portal for Grid-enabled molecular science.
  - Collaborative environments for microscopy and imaging.
  - Hydro-meteorological models in a distributed, remote environment.
  - A Grid environment for the analysis of MRI brain data.
- **Production Grids**
  Papers in this category mainly described large, functioning, Grid projects in Europe. One example was the DataGrid for the ATLAS/LHC project at CERN which will require 7 petabytes of data storage in the first year of its operation, in 2007. There was also a description of a lightweight Grid infrastructure which has been used for industrial applications.

20.2 The e-Science "Software Stack"

- **Grid Workflows**
  e-Science applications often require that different tasks need to be connected together by data flow and control flow. These workflows describe the throughput of scientific data through various algorithms, applications and services. The pathways through a given workflow are data dependent and resources dependent. Specific papers described:
  - Quality of Services (QoS) support for time-critical workflow applications.
  - Use of commercial workflow engines for scientific applications.
  - Algorithms for workflow scheduling on Grid resources.
  - Languages and models for scientific workflows.
- **Resource Management and Scheduling**
  This theme is related to workflows but with more of a focus on the Grid platforms themselves. Specific papers described:
  - Protocols for reservation and co-allocation of Grid resources.
  - Adaptive Grid reconfiguration.
  - Grid scheduling, optimization, resource management and task distribution.
  - Job submission issues.
- **Data Management**
  This theme dealt with data replication, transfer and management for Grid computing. Specific papers described:
  - Parallel file transfer protocol.
  - Input Output services for Grids; Data replication services.
  - GridFTP.
  - Prefetching of data for distributed computations.
- **Web Services and e-Science Enabling Technologies**
  This theme is similar to the Applications theme but with a greater emphasis on enabling technologies. Web Services are now the basis of Globus and the Open Grid Services Infrastructure (OGSI) but there are other homegrown frameworks as well. The Web Services technology stack includes XML, SOAP and WSDL. Specific papers described:
  - Architectures for trusted Web Services.
  - Case studies of Grid management systems using Web Services.
  - Binary XML for scientific data.
  - Reliable messaging in Web/Grid services environments.
  - An Agent-oriented framework for Grid computing.
  - Web Services orchestration using a functional language.
- **Service Management**
  Papers in this theme discussed the installation and management of Grid applications and the lifecycle management for Grid infrastructure itself.
- **Security**
  Specific papers in this theme discussed:
  - Access control for Grid services.

- Dynamic creation of Grid Virtual Organizations.
- Grid account management.
- Identity-based cryptography for Grid services.
- Simulations of Grid security services.
- **Peer to Peer - Networking**
  Papers in this category discussed rerouting strategies for networks and a semantics-based routing scheme for Grid resource discovery.
- **Workshop of Problem Solving Environments on Distributed Resources**
- **Workshop on Scientific Instruments and Sensors on the Grid**

20.3 Server-Side EScope for DataGrids (with Raju Karia)

EScope can be turned into a server-side software system to provide a browser-based portal to a DataGrid. In this section we will describe a prototype system, called "WebScope", which has been developed by students in the e-Science program at the Australian National University[3].

Apache Tomcat(http://tomcat.apache.org/) is an open-source Java servlet container which is used for a number of large, industrial-strength Web applications around the world. After downloading and installing Tomcat, the following steps are needed to convert a Java application, like EScope, to be accessible across the Web:

- The Java classes which you need to use must be rewritten as servlets.
- An information file, typically in XML format, needs to be written to direct Tomcat to open the first Java servlet when a connection is made.
- The entire set of servlet classes is often placed in a special directory to be read by Tomcat.

Once this work has been done, Tomcat will take care of managing concurrent requests for WebScope services over the internet.

Conceptually it is possible to think of EScope as having two parts:

1. The user interface, including graphics, will need to be displayed on a browser.
2. The data server software domain negotiates access to MDSplus and other databases.

A server-side version of EScope could build the user interface from a combination of web forms, made dynamic using Java Server Pages, and Java Applets. Applets are self-contained classes which get downloaded to the client browser and run by a local Java Runtime Environment. They have been a feature of Java since the early days and originally led to much of the enthusiasm

---

[3] We especially wish to acknowledge the efforts of Ajith Mose and Shi Hu for their work on projects supervised by Henry Gardner and Raju Karia.

about Java and the internet revolution. These days Applet technology is not as glamorous because:

- Java Runtime Environments no longer come as a default with major Web browsers.
- Applets are "heavyweight" and can be very slow to download.

Because of these issues, modern graphical Web applications are trending towards technologies which can function on any browser client and which exchange smaller parcels of data between the client and server. One of these is an amalgam of technologies called AJAX ("Asynchronous Javascript and XML") which enables the Web pages to be incrementally updated (rather than being reloaded statically) with an asynchronous exchange of small amounts of data between the client and server. Incidentally, there are Java interfaces to AJAX such as ECHO2 (http://www.nextapp.com/platform/echo2/echo/), Google Web Toolkit (http://code.google.com/webtoolkit/) or Wicket (http://wicket.sourceforge.net/. It is possible to use these together with Applets in WebScope.

On the data management side, there are a range of accessible Java technologies which can be used to start building a prototype DataGrid management system. These will also be described in the next section.

20.3.1 Metadata Indexing, Persistence and Provenance in WebScope

The data-server part of EScope can be ported to the Grid using some of the design patterns discussed in Chapters 17 and 18. As discussed there, and in the sections above, a porting to a Grid/Web Services environment will need to take into account the discovery of, and negotiation with, networked computing and data resources. But once scientific data has been discovered, and access to it has been negotiated, it will still need to be searched, downloaded and post-processed. In order to manage these phases, e-Science DataGrid servers will need to develop and maintain indexes of *metadata*. This is particularly the case with systems such as MDSplus which are not relational databases and, thus, will not be able to support sophisticated search queries without having an additional layer of metadata.

What will scientific metadata look like? It is often difficult to answer this question *a priori*. For example, an MDSplus database might need to be searched to return "all RFX shots taken in August 2006 with central densities greater than 10^{13}". But the central pressures needed to support this search would not be recorded by the diagnostic measurements on the experiment. Instead, they would be obtained by *post-processing* diagnostic data. Somehow this post-processing needs to be carried out and the original data then needs to be tagged in such a way that relational queries can be supported. Because of this, metadata needs to be created over the lifetime of the DataGrid and it needs to be persistent. A utility written for Java known as Hibernate (www.hibernate.org) can help to achieve this.

One possible architecture for a web-enabled version of **EScope** is shown schematically in Fig. 20.1. In this system, the **WebScope** DataGrid server maintains a relational table of metadata which is managed by Hibernate. Hibernate performs object-relational mapping, maintains object persistence between sessions and also maintains a dual-level system of data caches to enhance performance. The metadata tables could index datasets on their primary keys (experiment name, shot number and diagnostic name) as well as secondary keys determined by post-processing. Over time the metadata indexing into the actual **MDSplus** database could become quite complex, but the system of tables corresponding to this system would be managed by Hibernate. As well as storing metadata obtained from post-processing, a system such as **WebScope** could record dataset usage and cross-link to a database of trusted users. This cross-referencing might not only be used for accounting purposes but it could help to determine the "trustworthiness" of the data itself: if a dataset is frequently downloaded and used for post-processing by trusted users then it probably contains data which is reliable and which might be interesting for other potential users. (Note that the GridSphere environment, mentioned in Sec. 20.2 includes Hibernate.)

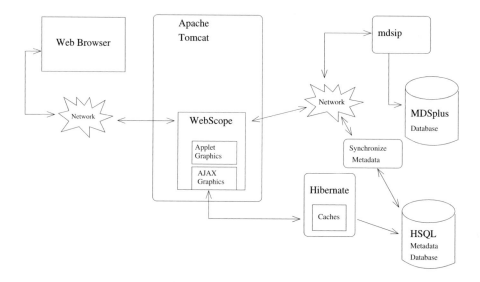

Fig. 20.1. Schematic diagram of the WebScope system described in the text.

Finally, as discussed above, our prototype **WebScope** system could include AJAX graphics and GUI components as well as Applets and traditional HTML forms.

20.4 A Final Word

We have tried to make the case that a study of the Java programming language together with an appropriate case study is a good entry point into the new discipline of e-Science. e-Science is about building software infrastructure for advanced, distributed, computational systems. Good software engineering practice is critical if this infrastructure is to last. Design patterns are one aspect of good software engineering practice.

MDSplus really does work. In spite of its more fiddly aspects, MDSplus is a highly-optimized, data-management system which is presently being used to store and access many Terabytes of scientific data. Although it is a product of fusion-energy research, it is a candidate for use in e-Science DataGrids in other areas and is particularly suited to the management and processing of waveform data.

EScope really does work. It is an interactive, networked, graphical waveform browser which can be used for scientific data from a wide variety of disciplines. All that is required is that a data server conforms to the DataFacadeInterface discussed in the text. Its interactive graphics are, we think, quite advanced and very useful for scientific users.

As we have said, design patterns really do work. The patterns that we have illustrated in this book have enabled us to build a software architecture for EScope which has been reasonably robust and extensible. This basic architecture should also be reusable for other similar software projects. Reusable architectures should greatly simplify the task of getting started on a software project. Eventually one would expect that they could be archived and made accessible to programmers in much the same way that libraries of algorithms are. The idea of constructing a large, authoritative "5 volume book" of software architectures was what led to the development of design patterns in the first place[4]. The authoritative book on architectures has not yet eventuated but design patterns have. As collective experience deepens, it is surely the case that reusable software architectures will be developed for e-Science as well as other areas of computing.

---

[4] Quoted from Erich Gamma's opening keynote speech to the ECOOP 2006 conference in Nantes, France, in July 2006.

A
Installing and Running Data Servers for EScope

It does not make a lot of sense to read this book without running the case study and other example software. In this appendix, we describe how you can go about setting up data servers, including MDSplus, which will work with versions of the PreEScope and EScope software supplied on the accompanying CD. In Sec. A.5, we also describe some of the basic tools supplied with MDSplus and give some background to setting up your own databases using MDSplus.

The data servers which we describe here range from "easy to install but slow to run" to "tricky to install but fast to run". MDSplus can be downloaded from its home web site. The web site accompanying this book should also be referred to for up-to-date installation information.

A.1 The MdsipSimulator Program

As explained in Sec. 2.7.1, *mdsip* is the communication protocol used by MDSplus. On the accompanying CD, we have supplied a Java program which simulates aspects of mdsip and which will work with two sample databases. This program, MdsipSimulator.java, is located in the source/mdsipSimulator directory. To use it, do the following:

1. Compile MdsipSimulator.java to produce two class files.
2. Move these class files into the examples directory. The two example databases, h1data_37025 and edam_17615, are located in this directory.
3. Run the simulator by typing java MdsipSimulator on a command line.
4. If all is well then you should be able to run any version of PreEScope or EScope in another command window. You will need to specify "localhost" as the server and the correct port number.
5. The port number can be seen at the end of the MdsipSimulator.java source code. If there is a problem running this code, then it is probably because another example program has grabbed port 8000. So try running it with another port number (say 8003).

6. For example, PreEScope0 can be run from another command window by typing:

```
java PreEScope0 localhost:8003 h1data 37025 .operations:i_fault
```

You should refer to the appropriate sections of the book for information about running the versions of PreEScope and EScope. For example, only the h1data database can be used for PreEScope0. Later versions of the case study can be used with either of the two sample databases. These are cut-down versions of the cut-down databases supplied for use with MDSplus! You will find that they can be slow to load into EScope and that some waveforms will be slow to display. These databases are, in fact, conventional directory trees of binary data. In contrast, MDSplus is highly optimized for storing huge amounts of waveform data. So it is worthwhile persevering with installing and running it when you have the time.

Good datasets to look at with these sample files are

- From h1data_37025 try

    .OPERATIONS.I_FAULT
    .RF.A14_4.INPUT_2

- From edam_17615 try

    .SIGNALS.PB.PBMC01_IM
    .SIGNALS.TP.TBMC01_IG

Some of these files can take a long time to download. Some others take even longer!

Also note that this simulation program does not distinguish between "child" and "member" nodes of the file tree as MDSplus does (see Sec. 6.5.1) and the directory trees for these sample databases will look slightly different to the MDSplus examples shown in the book.

A.2 The Text Data Server

This option is an alternative for those readers who are getting frustrated with the slow response of MdsipSimulator but who have not been able to get MDSplus running properly on their system.

Starting from EScope10, we introduce another data server which works with simple, columnar, text data. You can obtain the Java code for this server, TextDataServer.java, from the EScope10/textDataServer directory and integrate it with earlier versions of EScope from EScope4 onwards. Using the text data server will still enable all of the important patterns in this book to be illustrated and it will be extremely fast. But you will only be working with toy data.

A.3 Installing MDSplus

As with many other open-source projects, it is possible to download the MDSPlus source files as well as the binaries. The source code is useful when you need to develop new system components, or if you need to have a deeper control and understanding of the software organization. The binaries allow a quick system installation and we shall follow this approach here. It is possible to install the executable version both for Windows and Linux. Both the MDSPlus install shield for Windows and "RPM" for Linux can be downloaded from www.mdsplus.org. We have provided versions of both on the CD which also include a demonstration version of EScope and the demonstration datasets. These are, by far, the easiest way to get started.

A.3.1 Installing on Microsoft Windows Using the Install Shield on the CD

At the time of writing, the MDSplus.exe install shield supplied on the CD has been tested with WidowsXP. It appears that nothing could be easier! To use it, make sure that Java1.5 is already installed and then step through the following:

- Copy the install shield to your computer, perhaps to the desktop, and click on it.
- Follow the step-by-step instructions and choose the defaults unless you have a good reason not to. The main mdsip server will be installed under your Programs directory.
- To run the software, go to the Windows Start menu and choose All Programs – MDSplus – mdsip Data Server. The server will start listening to port 8000.
- Now try the version of EScope which is also available from the Start menu. This corresponds to EScope9 from Chapter 16.
- You can compile and run all of the versions of EScope and PreEScope supplied on the CD if they connect to port 8000 on localhost.
- Other programs such as jScope are also available from the Windows Start menu.
- See the accompanying web site for this book for up-to-date versions of the install shield and other information.

A.3.2 Installing on Windows from www.mdsplus.org

From www.mdsplus.org, click on "Software" in the left hand contents panel. Then click on "Downloads". You will be asked to submit your name and email address and a comment about your proposed usage of the software. Just say that you bought this book and want to play with MDSPlus! The authors of MDSPlus are nice people and are enthusiastic to have people using it. At the bottom of the form is a drop down menu to choose the installation platform. Choose "Windows".

At the time of writing, filling out this first form and submitting it results in you being asked to fill out an almost identical form again! This time you will be given a choice of "kits" to select for your platform. Our test installation was carried out using Windows XP. Select the appropriate kit and attempt to download the software.

You will now receive a professionally-constructed installation shield. Windows XP might complain about security but it did not do us any harm. Just proceed as you would do if you were downloading any other software package from the internet then execute `Setup.exe` and follow the prompts.

We chose all of the defaults for the location of `MDSplus` and other software from this download. It turns out that one of our sample databases needed access to more `MDSplus` functions then were included in the standard release. We have supplied an additional file, `Tdi.zip` which includes these functions. To install them, you copy the zip file to the `TDI` subdirectory of the main `MDSplus` install directory, for example:

```
C:\Program Files\MDSplus\TDI
```

Then unzip this file into this `TDI` subdirectory. `MDSplus` will find the functions when it needs them.

A.3.3 Installing on Linux Using the Supplied RPM

RPM stands for "Red-hat Package Manager" and it is meant to be an easy way of installing software. In truth, there is more fiddling about than for the Windows install shield. As for Windows, Java1.5 needs to be already installed if the provided RPM for `MDSplus` is to work.

Logon as root and type the following at a command line in the same directory as the supplied RPM file:

```
rpm -iv MDSplus-0.1-1.i386.rpm
```

You need to set your PATH environment variable using the following command (in bash)

```
export PATH=$PATH:/usr/local/mdsplus/bin
```

Then you can start `mdsip` via the command

```
start_mdsip
```

EScope can be started using the command `start_escope`, jTraverser can be started using the command `start_traverser`, jScope using the command `start_scope`.

The RPM installer checks also for library compatibility, so there may be some problems on some older Linux installations. If installation fails due to some failed "lib" dependencies, then you should attempt to install the relevant libraries.

A.4 Running MDSplus with the Sample Data

If you have installed MDSplus using the install shield from the CD then you do not need to set the following environment information. If not, then you need to read on.

Before running MDSplus, some attention needs to be put into ensuring that your computer has the correct environment path information.

To begin, we need to define an environment variable which tells MDSplus where it can obtain some functions it needs to evaluate expressions. The name of this environment variable is mds_path. On Linux (using the Bash shell) we define the environment variable as follows:

```
export mds_path = C:\Program Files\MDSplus\TDI
```

On Windows, things are a more complicated, since we need to define the variable in the registry. These are the required steps on Windows XP (other versions of Windows might vary slightly):

1. Start the registry editor program regedt32 from the Windows Run menu.
2. Open the HKEY_CURRENT_USER folder in the directory panel.
3. Open the Software folder.
4. If the folder MIT is not already present inside the Software folder, then create it with Edit -> New -> Key. This will create a new folder which you can rename to MIT.
5. In the folder MIT, if the folder MDSPLUS is not already present, create it (again, with Edit -> New -> Key).
6. In the folder MDSPLUS, create a mds_path variable name (Edit -> New -> String Value). You will see this name appear in the right hand panel.
7. Double click on the variable name in the right hand panel to open up a window called Edit String which lets you fill in the field Value data:. Fill this in with with the variable name and fill the field String: with the full path of the directory to contain the tree. (For example C:\Program Files\MDSplus\TDI)

Note that on Windows, the variable is defined forever, while on Linux you need to define it for each session. On Linux it is, therefore, convenient to put the definition in a start-up shell script, such as .bashrc.

Next, we need to define the appropriate environment variables for the sample databases. On the CD, the h1data database is contained in the file h1data_mdsplus_37025.zip (in the source directory). Unzipping this file will place the data files into a directory called h1data. The full path to this directory needs to be set as an environment variable with the name h1data_path. The procedure to do this is the same as for the mds_path environment variable. Similarly, the rfx_edam_mdsplus_17615.zip file will unzip into a directory called edam and the corresponding edam_path environment variable needs to be set.

Let us now run `MDSplus`! In a command window, change to the main `MDSplus` directory. (On Windows usually `Program Files/MDSplus`, on Unix usually `/usr/local/mdsplus/bin`). You should see a program, `mdsip.exe` in this directory. This is the program which serves `MDSplus` over the internet. Run it with the command:

```
mdsip -p 8000 -m -h mdsip.hosts
```

Here the flag `-p` indicates the port number which the `mdsip` server listens to and the flag `-h` indicates the file which defines the access rights for clients. (We shall use the default file provided with the `MDSplus` distribution.)

You can test that all is well by running any of the versions of `PreEScope` or `EScope` in another command window!

A.5 TCL, Traverser and Scope

This section describes some of the tools supplied with the `MDSplus` distribution and explains how they can be used to set up a simple, example database.

Once installed, several MDSplus tools will be available in the start menu for Windows, or callable from a terminal in Linux. We shall use the following tools to set up an example experiment database:

- Tree Command Language (TCL): a command interpreter for building experiment databases (Linux command: `mdstcl`);
- Traverser: a graphical interface to the experiment database (Linux command: `jTraverser`);
- Scope: the `MDSplus` waveform visualization tool (Linux command: `jScope`).

The Traverser and Scope applications were originally written in C together with native, windowing libraries. They have Java versions, `jTraverser` and `jScope` which are available from the `MDSPlus` download and which we will use in the examples described here.

We shall now build a very simple, tree-structured experiment database containing several types of data. A data item may be a number, a string, a signal or, more generally, an expression i.e. a combination of data and operators.

A.5.1 Creating a Simple Database Using TCL

The first step is to set up an environment variable called `my_tree_path`. This will provide the address of a directory called `my_tree`.

We can now create now a database containing three data items:

NUM1 containing a number
NUM2 containing an array
NUM3 containing an expression

TXT containing a string

We can achieve this by the following steps:

1. Start TCL (from the `Start -> mdsplus` menu on Windows, or using `mdstcl` on Linux)
2. Create a new tree:
   `TCL>edit my_tree/new`
3. Add node NUM1:
   `TCL>add node NUM1/usage=numeric`
   This command creates a new node named NUM1. It will contain numeric data.
4. Fill node `NUM1`:
   `TCL>put NUM1 "2"`
   This command fills node NUM1 with the number 2.
5. Add node NUM2:
   `TCL>add node NUM2/usage=numeric`
6. Fill node NUM2:
   `TCL>put NUM2 "[1,2,3,4,5,6,7]"`
7. Add node NUM3:
   `TCL>add node NUM3/usage=numeric`
8. Fill node NUM3:
   `TCL>put NUM3 "NUM1 + 3 * NUM2"`
   Node NUM3 now contains an expression involving the contents of nodes NUM1 and NUM2.
9. Add node TXT:
   `TCL>add node TXT/usage=text`
   This command creates a new node named TXT. It will contain text.
10. Fill node TXT:
    `TCL>put TXT " 'This is a text string' "`
11. Write the current tree to disk:
    `TCL>write`
12. Close the tree:
    `TCL>close`

It is worth noting that the values we have just inserted into the nodes of `my_tree` represent quite different things: the number 2, an array of integer values, the string "This is a text string". Nevertheless, within `MDSplus`, they all represent expressions. The "expression" is a central concept in `MDSplus`: every entry is an expression. An expression can be something as simple as a number or a node reference, but it may also be a very long combination of numbers, references and operators. For example, the expression NUM1 + 3 * NUM2 defining the content of node NUM3 evaluates to 2 + 3*[1,2,3,4,5,6,7]=[5,8,11,14,17,20,23]. Note that what is stored in the tree is the expression definition, not its evaluated value. Expression evaluation is done on-the-fly every time that node is retrieved.

What we have just created is a tree-structured database. If you look at the files created in the directory specified by my_tree_path, you will see that the database is composed of three files: my_tree_model.characteristics, my_tree_model.datafile and my_tree_model.tree. It is possible to insert, modify and retrieve the contents of these data nodes. However, in many experimental situations we are often concerned with storing data once as it is acquired rather than modifying a particular database over time. For example, every "shot" of a fusion experiment would populate a separate instance of a MDSplus database with data. To accomplish this, MDSplus is first used to create a template database (called an "experiment model"). Before a shot is run, this template database contains the experimental set-up values, as well as defining the places (represented by empty nodes in the tree) where acquired data will be stored. This template database gets replicated for each shot, and is filled with the data acquired during that shot.

A.5.2 Examining a Database Using the Traverser

We are now ready to look at our newly created database using the Java version of the "Traverser" tool:

1. Start the Traverser tool (Linux command: jTraverser);
2. Give it the command File -> Open
3. Write my_tree in the Tree: field and -1 in the Shot: field of the pop-up window.

Fig. A.1. A graphical view of our experiment model.

Now we have a graphical view of our tree as in Fig. A.1 You can see the names of the nodes that have just been created and the icons indicate the "usage" for that node (numeric or text etc.). When data is inserted into a node,

A.5 TCL, Traverser and Scope

the data access layer of MDSplus checks whether its type matches the specified usage for that node. Using jTraverser and pressing "Mouse_Button_3" (MB3) over a node pops up a menu which lets you perform several operations on the node content as shown in Fig. A.2.

Fig. A.2. The node-context pop-up menu.

The commands we are interested in are:

- Display Data: displays the content of the selected node;
- Display NCI: displays information such as size of the data and the insertion date;
- Modify Data: modify the content of the node;

When modifying data for node NODE3, jTraverser displays the dialog shown in Fig. A.3. You can type any expression which replaces the current content of the node. To experience the check performed on the node usage when inserting data, try to change the content of the node NUM3 to 'This another string' and to save it. You will receive the error message fisted in Fig. A.4:

This error message indicates that the usage of that node is not correct – it must be an expression returning a numeric value.

A.5.3 Creating and Viewing Subtrees

Up to now, we have created a flat collection of nodes, possibly containing data. Let's now exploit the hierarchical database organization by adding a subtree called SUB1 containing a numeric node SUB_NODE1 and another subtree called SUB2, containing a text node SUB_NODE2. We shall again use the Tree Command Interpreter as follows:

TCL>edit my_tree Open tree my_tree for editing (i.e. adding/removing nodes). The default shot number is -1 (the experiment model)

236   A Installing and Running Data Servers for ESCope

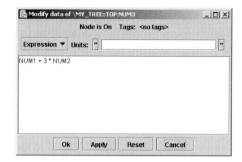

Fig. A.3. The modify data pop-up dialog.

Fig. A.4. The message box indicating an attempt to define a wrong data type.

TCL>add node .SUB1 Creates a new subtree. Note that the name is preceded by a dot.
TCL>set def .SUB1 Moves into SUB1 subtree. Much like the UNIX cd command.
TCL>add node SUB_NODE1/usage=numeric Add empty node SUB_NODE1 to subtree SUB1.
TCL>add node.SUB2 Creates subtree SUB2.
TCL>set def .SUB2 Move into SUB2 subtree.
TCL>add node SUB_NODE2/usage=text Add empty node SUB_NODE2 to subtree SUB1 to contain a text string.
TCL>write Write the newly created tree.
TCL>close Close the tree.

Now, when we open my_tree with jTraverser, and expand the subtrees, we get the the tree structure shown in Fig. A.5 (you can expand and contract subtrees by clicking on the associated handles or by double clicking the subtree):

A.5.4 Understanding Node Names

In a tree data structure, it is necessary to define the whole path to a node in order to give it a unique name. Let's take an example: select the pop-up item *show data* in jTraverser over the node SUB_NODE2 and stretch the window so that the whole title becomes visible (Fig. A.6)

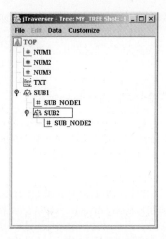

Fig. A.5. Graphical view of the modified database hierarchy.

Fig. A.6. A display data dialog which shows the full path to a node.

The dialog tells us that the node is undefined because it does not contain data yet, but we are only interested in its title for now. This shows the path name of node SUB_NODE2. The first part \MY_TREE::TOP stands for the root of tree MY_TREE and the rest of the name is the path from the root to node SUB_NODE2. Observe the dots and the colons. The hierarchy organization in MDSplus trees defines two kinds of nodes for every subtree: "members", whose name is preceded by a colon, and "children", whose name is preceded by a dot. In this example SUB_NODE2 is a member of node SUB2 which is in turn a child of node SUB1. The MDSplus data organization defines also the concept of default position. Node pathnames can in fact be absolute (i.e. starting from the root) or relative, i.e. starting from a default node. Even though MDSplus allows an arbitrary organization of members and children (a member node may have members and/or children), the usual approach is to define members for containing data and children for defining the structure (data can never inserted into a child node).

This simple example should convince us that node pathnames can be very lengthy, increasing also the probability of typing errors. For this reason, MDSplus allows one or more unique names to be associated with a given node. These identifiers are called **tags** and they are very useful for providing a unique name for particularly important nodes. It is possible to define tags in TCL

Fig. A.7. The add/remove tags pop-up window.

but it is easier to do it with jTraverser. To give tag name MY_SPECIAL_NODE to node \MY_TREE::TOP.SUB1.SUB2:SUB_NODE2 using jTraverser, you open experiment my_tree selecting also the *edit* checkbox in the open dialog. Then you position the mouse over the node, press MB3 and select "Modify Tags" in the pop-up menu. You then get the dialog of Fig. A.7 in which you can add/remove the tag name(s) which will be associated with that node. Add MY_SPECIAL_NODE in the list, writing the tag name in the *Current Selection* field, pressing the *Add Tag* button and finally the *Ok* button. From now, the tag is associated with that node,which can now be referred either by its path name or its tag. Tag names can be used everywhere a node reference is required (for example, in an expression referring to that node) and the general syntax is

\<tree name>::<tag name>

In our example this would be:

\MY_TREE::MY_SPECIAL_NODE

When only one tree is open (as in the present example) the first part, <tree_name>:: can be omitted (retaining the backslash). Tag names can have an arbitrary number of characters. (But node names are limited to 12 characters.)

Let's now change the content of a node, using the *Modify Data* option in the node-context pop-up menu of Fig. A.3, activated over node SUB_NODE_1. The first time we shall get the dialog shown in Fig. A.6: the node doesn't contain data, as no data has been added when we created it using TCL. If we select the *Expression* option in the menu, we can then type any valid expression for this node, such as \MY_SPECIAL_NODE * 4.

A.5.5 Defining Signals and Viewing Them with jScope

In our examples we have so far dealt with strings, numbers, arrays and expressions. Another data type which is very useful in data acquisition is the *signal* which is an array of data (y axis values) with associated x axis information. Most times, signals refer to some physical signal acquired at a certain frequency for a certain time. In this case the X axis represents the time value of the acquired samples. MDSplus defines an explicit data type for signals and a signal "usage" for tree nodes. Though signals are usually produced by some data acquisition routine, it is possible to define signals manually by writing the corresponding expression. We shall not the detailed syntax for signals which is part of the fairly complicated MDSplus syntax for describing expressions. Rather, we shall illustrate it by means of another simple example.

Let's add a node to my_tree called SIGNAL1 containing a sine wave made of 1000 points sampled over a period of 1 second.

TCL>edit my_tree Open my_tree in edit mode.
TCL>add node SIGNAL1/usage=signal Add node SIGNAL1, designed to contain a signal.
TCL>put SIGNAL1 "build_signal(...)" With the argument of build_signal equal to sin(6.28 *[0..999]/1000.), , [0..999]/1000.), this will write a sinusoidal waveform in node SIGNAL1
TCL>write Write everything on disk.
TCL>close Close the database.

In this example, the syntax for defining signals is the following:

BUILD_SIGNAL(DATA, RAW, [DIMENSION...])

where:

DATA is an expression defining the Y values. (sin(6.28 * [0..999]/1000.) in our example)
RAW is an expression indicating raw data. Often an acquired signal is made of raw data then converted by taking into account parameters such as gain and offset. In this case we are not interested in this subtlety so we omit its definition and replace RAW with the two commas in the TCL example.
DIMENSION is an expression returning the array of X axis, usually (but not always) time values. (In our example the expression is [0..999]/1000.)

We are now ready to use jScope to look at the waveform we have just inserted in the database. Before doing it, let's assign a tag name to the newly created node SIGNAL1. Do this using the traverser tool for assigning the tag SINE_SIGNAL to the node SIGNAL1 following the example in the previous section.

To run jScope on a Windows system, it suffices to double click on the Scope icon. On Linux, the command is jScope. When started, jScope displays an empty panel. To connect it to the desired waveform (the content of node SIGNAL1 in our case) it is first necessary to select local disk access as the data source selector, by selecting the Local server in the server list at Network->Servers, as shown in Fig. A.8. We are, in fact, now using jScope to display data stored on a local disk. Afterwards, it is necessary to connect the panel to the signal by filling the pop-up setup window, activated via

240   A Installing and Running Data Servers for ESCope

Fig. A.8. Setting disk data as the data server for jScope.

the pop-up menu item Setup Data Source..., as shown in Fig. A.9 (recall that we are now working on the experiment *model*, whose shot number is, by convention, -1).

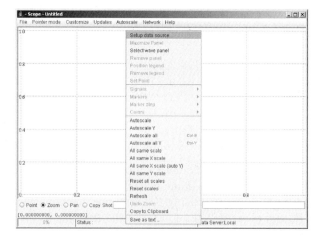

Fig. A.9. The pop-up option list activated by pressing the right mouse button over a jScope panel.

At this point, the sinusoidal waveform is displayed, and we can play with it, e.g. by zooming parts of it, or moving a cross-hair cursor and looking at the corresponding typed x and y values. Let's now become confident with expressions in jScope and create one or more additional panels us-

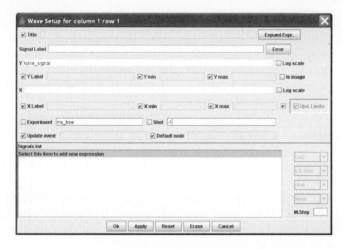

Fig. A.10. Connecting the panel to the waveform in jScope.

ing `Customize->Windows` option in the menu bar. If we "connect" a panel (again, selecting *Setup data Source* item in the pop-up menu) to expression `\SINE_SIGNAL * NUM1` we shall see the sinusoidal waveform multiplied by 2, i.e. the value contained in node NUM1. (Note that the default position in a tree is the root, so we can avoid writing the whole pathname `\MY_TREE::TOP.NUM1` to specify the path of node NUM1.)

If we connect a `jScope` panel to NUM2 containing the array [1,2,3,4,5,6,7], we shall see a ramp whose Y values range from 1 to 7. What about X axis? Unlike signals, in fact, arrays do not bring information about X axis, and therefore jScope assumes a sequence of integer values as X axis, corresponding to the indices of the array values.

As a final exercise we will show how a sinusoidal waveform can be defined using a more complicated expression syntax. We can first replace the contents of NUM1 by the expression `SIN(6.28 * [0:999]/1000.)` and then put the expression `[0:999]/1000.` into node NUM2. Now we can connect a `jScope` panel to the expression `BUILD_SIGNAL(NUM1,,NUM2)` and shall the sinusoidal waveform again. Alternatively, we could have have stored the expression `BUILD_SIGNAL(NUM1,,NUM2)` in node `SIGNAL1` and just viewed this node using `jScope`.

A.5.6 UNITS_OF() and DIM_OF()

The syntax for expressions in MDSplus is very rich and they can include many primitive functions which you can read about at the `MDSplus` website. Two of these primitive functions, `DIM_OF()` and `UNITS_OF()` will be used later on in `EScope` to interact with `MDSplus` data servers so we will extend the previous example to illustrate their use.

The function `DIM_OF()` returns the X axis data of the signal passed as argument. If we open the *Setup Data* dialog in `jScope` and substitute the Y axis expression `\SINE_SIGNAL` with `DIM_OF(\SINE_SIGNAL)` we see a ramp instead of the sine waveform. In fact, the expression `DIM_OF(\SINE_SIGNAL)` returns the X axis of `\SINE_SIGNAL` which is an array of 1000 increasing values for the X axis ranging from 0 to 0.999.

The function `UNITS_OF()` returns a string which describes the "units" of a given expression (if it has any units). Probably you already observed that the dialog in `jTraverser` for writing expressions also has a *Units* field. You can type a string there, which will be associated with the expression. Try writing the string "Volts" in the *Units* field for node NUM1 and then fill the *Title* field in the *Setup Data source* dialog with the expression `UNITS_OF(NUM1)`. You will see string "Volts" written at the top of the panel: `jScope` evaluates the expression defined in the *Title* field of a node and displays the returned string value as a title in the corresponding panel.

B
Listings of Introductory Examples

B.1 BorderComponentFrame

```java
import java.awt.BorderLayout;
import javax.swing.*;
/** Places a number of Swing GUI components onto a frame
    using border layout.*/
public class BorderComponentFrame extends JFrame
{
    private JPanel componentPanel;
    public BorderComponentFrame()
    {
        //Components for the GUI
        JMenuBar menuBar = new JMenuBar();
        JMenu menu = new JMenu("Drop Down Menu");
        JMenuItem item = new JMenuItem("Menu Item");
        JButton aButton = new JButton("a button");
        JButton aNButton = new JButton("another button");
        JButton yANButton =
                    new JButton("yet another button");
        JLabel aLabel = new JLabel("label");
        JTextArea text = new JTextArea("Type here!");
        menuBar.add(menu);
        menu.add(item);
        componentPanel = new JPanel();
        componentPanel.setLayout(new BorderLayout());
        componentPanel.add(menuBar,"North");
        componentPanel.add(aButton,"East");
        componentPanel.add(aLabel,"South");
        componentPanel.add(aNButton,"West");
        componentPanel.add(text,"Center");
    }
    public void initialise()
    {
```

```java
            getContentPane().add(componentPanel);

            setSize(600,400);
            setLocation(200,200);
            setDefaultCloseOperation(JFrame.EXIT_ON_CLOSE);
            setTitle( "EScope: Border Component Frame" );
            setVisible(true);
    }
    public static void main(String[] args)
    {
            BorderComponentFrame ourEScope =
                                    new BorderComponentFrame();
            ourEScope.initialise();
    }
}
```

B.2 Plotter

```java
import java.awt.*;
import java.awt.event.*;
import java.awt.geom.*;
import javax.swing.*;
/** Supplies a static method to plot an array of x values
    against an array of y values in a separate frame.*/
public class Plotter
{
    public static void plot(double[] xVals, double[] yVals,
                                                String title)
    {
        DrawFrame f = new DrawFrame();
        f.initialise(xVals,yVals,title);
        f.setDefaultCloseOperation(JFrame.EXIT_ON_CLOSE);
        f.setVisible(true);
    }
}
/** A frame that contains a panel with drawings */
class DrawFrame extends JFrame
{
    public static final int WIDTH = 700;
    public static final int HEIGHT = 500;

    public DrawFrame(){}

    public void initialise(double[] xVals, double[] yVals,
                                                String title)
    {
```

```java
            setTitle(title);
            setSize(WIDTH,HEIGHT);
            DrawPanel panel = new DrawPanel();
            panel.setGraphData(xVals,yVals);
            getContentPane().add(panel);
    }
}
/** A panel that displays a line joining data points*/
class DrawPanel extends JPanel
{
    private double[] xVals,yVals;
    public DrawPanel(){}
    public void setGraphData(double[] x,double[] y)
    {
        this.xVals = x;
        this.yVals = y;
    }
    public void paintComponent(Graphics g)
    {
      BasicStroke bs;                           // Ref to BasicStroke
      int i;                                    // Loop index
      Line2D line;                              // Ref to line
      float[] solid = {12.0f,0.0f};             // Solid line style
      Graphics2D g2 = (Graphics2D) g;           // Cast to Graphics2D
      Dimension size = getSize();               // Get plot size
      // Set background color
      g2.setColor( Color.white );
      g2.fill(new Rectangle2D.Double(0,0,size.width,
                                     size.height));
      // Set the Color and BasicStroke
      g2.setColor(Color.black);
      float strokeWidth = 1.0f;
      bs = new BasicStroke( strokeWidth,
                 BasicStroke.CAP_SQUARE,
                 BasicStroke.JOIN_MITER, 1.0f, solid, 0.0f );
      g2.setStroke(bs);
      double xMax,yMax,xMin,yMin,deltaX,deltaY;
      xMax=xVals[0]; xMin=xMax;
      yMax=yVals[0]; yMin=yMax;
      for (i=1;i<xVals.length; i++)
      {
          if (xVals[i] > xMax) xMax = xVals[i];
          if (yVals[i] > yMax) yMax = yVals[i];
          if (xVals[i] < xMin) xMin = xVals[i];
          if (yVals[i] < yMin) yMin = yVals[i];
      }
      deltaX = xMax - xMin;
      deltaY = yMax - yMin;
      double xScale,yScale;
```

```
            xScale = (size.width)/(deltaX);
            yScale = (size.height)/(deltaY);
            double[] xScaled = new double[xVals.length];
            double[] yScaled = new double[yVals.length];
            for (i=0; i<xVals.length; i++)
            {
                xScaled[i] = (xVals[i]-xMin)*xScale;
                yScaled[i] = size.height - (yVals[i]-yMin)*yScale;
            }
            // Plot curve
            for ( i = 0; i < xVals.length -1; i++ )
            {
                line = new Line2D.Double(xScaled[i], yScaled[i],
                                    xScaled[i+1],yScaled[i+1]);
                g2.draw(line);
            }
        }
    }
}
```

B.3 ShotDataCache2

```
import java.awt.*;
import java.awt.event.*;
import javax.swing.*;
import javax.swing.event.*;
/**
    This program shows a list made up of a DefaultListModel
    which contains ShotData objects.
*/
public class ShotDataCache2
{
    public static void main(String[] args)
    {
        JFrame frame = new ShotDataFrame2();
        frame.setDefaultCloseOperation(JFrame.EXIT_ON_CLOSE);
        frame.show();
    }
}
/** Demonstrates the operation of a linked list */
class ShotDataFrame2 extends JFrame
{
    private static final int WIDTH = 400;
    private static final int HEIGHT = 300;
    private JList shotList=null;
    private DefaultListModel lModel=null;
    public ShotDataFrame2()
```

```java
{
    setTitle("ListTest");
    setSize(WIDTH, HEIGHT);
    lModel = new DefaultListModel();
    int shotNo = 37025;
    ReadPlotData reader = new ReadPlotData(
                    "DoperationsCi_fault", shotNo);
    ShotData data1 = reader.readData();
    reader = new ReadPlotData("DoperationsCdiamag", shotNo);
    ShotData data2 = reader.readData();
    reader = new ReadPlotData("DrfCrf_drive", shotNo);
    ShotData data3 = reader.readData();
    lModel.addElement(data3);
    lModel.addElement(data2);
    lModel.addElement(data1);
    shotList = new JList(lModel);
    shotList.setSelectionMode(
                ListSelectionModel.SINGLE_SELECTION);
    shotList.addListSelectionListener(new ShotListener());
    JScrollPane scrollPane = new JScrollPane(shotList);
    JPanel p = new JPanel();
    p.add(scrollPane);
    Container contentPane = getContentPane();
    contentPane.add(p, BorderLayout.SOUTH);
}
private class ShotListener
        implements ListSelectionListener
{
    public ShotListener() {}
    public void valueChanged(ListSelectionEvent event)
    {   //You get 2 events if you do not have the
        //following if!
        if (event.getValueIsAdjusting() == false)
        {   // (There is only one element in this list)
            Object[] values = shotList.getSelectedValues();
            for (int i = 0; i < values.length; i++)
            {
                ShotData shot = (ShotData)values[i];
                lModel.removeElement(shot);
                Plotter3.plot(shot.getXVals(),
                        shot.getYVals(), shot.getXUnits(),
                        shot.getYUnits(), shot.getTitle());
            }
        }
    }
}
}
```

C
Helper Classes for Accessing MDSplus from Java

The following listings are some of the the helper classes discussed for accessing MDSplus.

C.1 MDSDescriptor

```java
/** Used to store the response from the MDSPlus server */
public class MDSDescriptor
{
    public static final byte MAX_DIM       = 8;
    public static final byte DTYPE_CSTRING = 14;
    public static final byte DTYPE_CHAR    = 6;
    public static final byte DTYPE_BYTE    = 2;
    public static final byte DTYPE_SHORT   = 7;
                                        //converted to int
    public static final byte DTYPE_INT     = 8;
    public static final byte DTYPE_FLOAT   = 10;
                                        //converted to double
    public static final byte DTYPE_DOUBLE  = 11;
    public static final byte DTYPE_WORDU   = 3;
                    //"usigned word"; converted to int
    public static final byte DTYPE_EVENT   = 99;

    private byte   descriptorType;
    private byte   byteData [];
    private int    intData [];
    private double doubleData [];
    private String charData;
    private String cstringData;
    private String eventData;

    // Methods to set and get elemental values
```

```java
public void setByteDataElement
                            (byte i_byteData, int index)
{
    if (index>=0 && index<byteData.length)
        byteData[index] = i_byteData;
}
public byte getByteDataElement(int index)
{
    if (index>=0 && index<byteData.length)
        return byteData[index];
    return byteData[0];    // may want to change this
}
public void setIntDataElement(int i_intData, int index)
{
    if (index>=0 && index<intData.length)
        intData[index] = i_intData;
}

public int getIntDataElement(int index)
{
    if (index>=0 && index<intData.length)
        return intData[index];
    return intData[0];    // may want to change this
}
public void setDoubleDataElement
                            (double i_doubleData, int index)
{
    if (index>=0 && index<doubleData.length)
        doubleData[index] = i_doubleData;
}
public double getDoubleDataElement(int index)
{
    if (index>=0 && index<doubleData.length)
        return doubleData[index];
    return doubleData[0];    // may want to change this
}

// More set and get methods.
public void setDtype(byte _dtype)
                            { descriptorType = _dtype;}
public byte getDtype(){ return descriptorType;}
public void setByteData(byte[] _byteData)
                            { byteData = _byteData;}
public byte[] getByteData(){ return byteData;}
public void setIntData(int[] _intData)
                            { intData = _intData;}
public int[] getIntData(){ return intData;}
public void setDoubleData(double[] _doubleData)
                            { doubleData = _doubleData;}
```

```java
    public double[] getDoubleData(){ return doubleData;}
    public void setCharData(String _charData)
                                { charData = _charData;}
    public String getCharData(){ return charData;}
    public void setCstringData(String _cstringData)
                                { cstringData = _cstringData;}
    public String getCstringData()    { return cstringData;}
    public void setEventData(String _eventData)
                                { eventData = _eventData;}
    public String getEventData()      { return eventData;}
}
```

C.2 MDSDataSource

```java
import java.io.*;
import javax.swing.JTree;
/** This interface defines a Java communications
interface to the MDSPlus server.*/
public interface MDSDataSource
{
    /** Connect to the remote MDS server.
        - Check that the source is not already connected
        - Parse "source" to ensure that a port number is
          specified
        - Send JAVA_USER message
        @param serverAddrCPort The server address
                             and port separated by a colon
        @throws IOException */
    public void connect(String serverAddrCPort )
                                        throws IOException;
    /** Close the socket, set stream handles to null
       and connected flags to false.
       @throws IOException */
    public void disconnect() throws IOException;
    /** Open the socket by sending message
         "JAVAOPEN(experiment, shot)"
        - If open is not successful then the
          Descriptor returned by the
          server is null. Otherwise the open is successful.
        - Check that we are connected but not already open.
        - Set flags
    @throws IOException */
    public void open(String experiment, int shot)
                                        throws IOException;
    /** Send message "JAVACLOSE(experiment, shot)"
        - Set flags
```

```java
        @throws IOException */
    public void close() throws IOException;
    /** Evaluate an expression by sending a string
            to the server.
            If not null, the result will be contained
            in the data field of the
            MDSDescriptor returned by MDSPlus.
            - Check that we are connected and open
            @throws IOException
            @return MDSDescriptor returned by MDSPlus
    */
    public MDSDescriptor evaluate(String expression)
                                            throws IOException;
    public void constructTree(JTree tree) throws IOException;
    // get and set and get-status methods
    public boolean isConnected();
    public boolean isOpen();
    public String getExperiment();
    public int getShot();
}
```

C.3 MDSNetworkSource

```java
import java.io.*;
import java.net.*;
import java.util.*;
import javax.swing.*;
public class MDSNetworkSource implements MDSDataSource
{
    private String serverAddrCPort;
    private String experiment;
    private int shot;
    private boolean isConnected;
    private boolean isOpen;
    private Socket socket;
                            // socket for connecting to MDSPlus
    private DataOutputStream output;
                            // output stream for socket
    private DataInputStream input;
                            // input stream for socket
    public MDSNetworkSource()
    {
        serverAddrCPort = null;
        experiment = null;
        shot = -1;
        isConnected = false;
```

C.3 MDSNetworkSource

```java
        isOpen = false;
        socket = null;
        input = null;
        output = null;
    }
    /** Attempts to open a new connection to the MDSPlus
        server. If no port given the default is port 8000.
        Sets flags
        @param _source The address of the MDSPlus server and
        port
        @throws IOException If already connected */
    public void connect(String _source) throws IOException
    {
        serverAddrCPort = _source;
        if ( isConnected )
        {
            throw new IOException( "Already connected" );
        }
        /* First check if a port is specified in the
           string */
        int i = serverAddrCPort.indexOf(":");
        String addr;
        int port;
        if ( i == -1 )
        {
            addr = serverAddrCPort;
            port = 8000;
        }
        else
        {
            addr = serverAddrCPort.substring( 0, i );
            port = Integer.parseInt(
                serverAddrCPort.substring(
                            i+1, serverAddrCPort.length()));
        }
        /* Connect to remote server */
        socket = new Socket( addr, port );
        output = new DataOutputStream( new
                    BufferedOutputStream(
                            socket.getOutputStream() ) );
        input = new DataInputStream( new
                    BufferedInputStream(
                            socket.getInputStream() ) );
        /* Send login name */
        MDSMessage message = new MDSMessage( "JAVA_USER" );
        message.send( output );
        message.receive( input );
        /* Flag that we are now connected */
        isConnected = true;
```

```
    }
    /** Disconnects from the MDSPlus server, and resets
        flags.
        @throws IOException If not already connected */
    public void disconnect() throws IOException
    {
        if ( !isConnected )
        {
            throw new IOException( "Not connected" );
        }

        socket.close();
        socket = null;
        output = null;
        input = null;
        isConnected = false;
        isOpen = false;
    }
    /** @throws IOException If not connected, already open,
        or if the server is down or not responding */
    public void open(String _experiment, int _shot)
                                              throws IOException
    {
        MDSDescriptor status;
        if ( !isConnected )
        {
            throw new IOException( "Not connected" );
        }
        if ( isOpen )
        {
            throw new IOException( "Already open" );
        }
        status = sendMessage( "JavaOpen(\"" + _experiment +
                              "\"," + _shot + ")" );
        if ( status == null )
        {
            throw new IOException(
                      "Null response from server" );
        }
        else
        {
            if ( status.getIntData() != null &&
                         status.getIntData().length > 0 )
            {   //diagnostic write of return data from server
                System.out.println( "MDSNetworkSource::Open:"
                    + "result = " + status.getIntData()[0] );
            }
            experiment = _experiment;
            shot = _shot;
```

```java
        isOpen = true;
    }
}
/** @throws IOException if not connected to a server
            or no experiment open. */
public void close() throws IOException
{
    MDSDescriptor status;
    if ( !isConnected )
    {
        throw new IOException( "Not connected" );
    }
    if ( !isOpen )
    {
        throw new IOException( "Not open" );
    }
    status = sendMessage( "JavaClose(\"" + experiment +
                            "\"," + shot + ")" );
    isOpen = false;
}

/** Sends a message to the MDSPlus server
    and returns the response.
This is a high level method and calls the more
complicated sendMessage() method which hides the more
complex implementation necessary to communicate
with an MDSPlus server.
@param expression The message to be sent to the server
@return The response from the server
@throws IOException If not connected to a server or
 no experiment open.
*/
public MDSDescriptor evaluate(String expression)
                                    throws IOException
{
    if ( !isConnected )
    {
        throw new IOException( "Not connected" );
    }
    if ( !isOpen )
    {
        throw new IOException( "Not open" );
    }
    return sendMessage( expression );
}
/** Sends a message to the MDSPlus server and
  returns the response. This method works at a lower
  level than evaluate() and requires a MDSMessage object.
@param expression The message to be sent to the server
```

C Helper Classes for Accessing MDSplus from Java

```java
    @return The response from the server
    @throws IOException passed up from MDSMessage's
                                    send or receive methods
    */
    private MDSDescriptor sendMessage( String msg )
                                          throws IOException
    {
        MDSMessage message = new MDSMessage( msg );
        message.send( output );
        message.receive( input );
        return message.toDescriptor();
    }
    // get and set and status methods
    public boolean isConnected(){ return isConnected;}
    public boolean isOpen(){ return isOpen;}
    public String getSource(){ return serverAddrCPort;}
    public String getExperiment(){ return experiment;}
    public int getShot(){ return shot;}
}
```

C.4 MDSMessage

```java
import java.io.*;
import java.util.zip.*;
/** Provides low-level communication with the MDSPlus server.
    Note for students: Because this class is low-level it is
    "ugly". You do not need to understand the details.*/
public class MDSMessage
{
    private static int msgid = 0;
    /* Constants for bit-masking. */
    public static final int SUPPORTS_COMPRESSION = 0x8000;
    public static final byte SENDCAPABILITIES = 0x0F;
    public static final byte COMPRESSED = 0x20;
    public static final byte BIG_ENDIAN_MASK = (byte)0x80;
    public static final byte SWAP_ENDIAN_ON_SERVER_MASK =
                    0x40;
    public static final byte JAVA_CLIENT = 3 |
            BIG_ENDIAN_MASK | SWAP_ENDIAN_ON_SERVER_MASK;
    /* Other constants */
    public static final String EVENTASTREQUEST =
                        "----EVENTAST----REQUEST----";
    public static final String EVENTCANREQUEST =
                        "----EVENTCAN----REQUEST----";
    private int msglen;
    private int status;
```

C.4 MDSMessage

```java
    private short length;
    private byte nargs;
    private byte descr_idx;
    private byte message_id;
    private byte dtype;

    private byte client_type;
    private byte ndims;
    private int dims[];
    private byte body[]; // body is a byte array
    private boolean compressed = false;
                        //Is the file compressed?
/* Constructs an MDSMessage object containing
   a "message" string which will get sent to
   the MDSplus server.
   The protocol is to send this message and
   then to receive the output
   data from MDSplus.
*/
public MDSMessage( String expr )
{
    int i;
    status = 0;
    length = (short)expr.length();
    nargs = 1;
    descr_idx = 0;
    ndims = 0;
    dims = new int[MDSDescriptor.MAX_DIM];
    for ( i = 0; i < MDSDescriptor.MAX_DIM; i++ )
    {
        dims[i] = 0;
    }
    dtype = MDSDescriptor.DTYPE_CSTRING;
    client_type = JAVA_CLIENT;
                    //Java client is "big-endian"
    body = expr.getBytes();
}
/* Send the message contained by this MDSMessage
   object */
public void send( DataOutputStream s ) throws IOException
{
    int i;
    msglen = 48 + body.length;
    s.writeInt( msglen );
    s.writeInt( status );
    s.writeShort( (int)length );
    s.writeByte( nargs );
    s.writeByte( descr_idx );
    s.writeByte( msgid++ );
```

258   C Helper Classes for Accessing MDSplus from Java

```java
            s.writeByte( dtype );
            s.writeByte( client_type );
            s.writeByte( ndims );
            for ( i = 0; i < MDSDescriptor.MAX_DIM; i ++ )
            {
                s.writeInt( dims[i] );
            }
            s.write( body, 0, length );
            s.flush();
            message_id ++;
        }
        /* Receive the response from the MDSplus server.
           Decode and store
           the header of this message */
        public void receive( DataInputStream s )
                                            throws IOException
        {
            byte header[] =
                        new byte[16 + 4 * MDSDescriptor.MAX_DIM];
            int i;
            s.readFully(header);
            client_type = header[14];
            compressed = ((client_type & COMPRESSED) ==
                                                    COMPRESSED );
            msglen = byteArrayToInt( header, 0);
            status = byteArrayToInt( header, 4);
            length = byteArrayToShort( header, 8);
            nargs      = header[10];
            descr_idx = header[11];
            message_id = header[12];
            dtype = header[13];
            client_type = header[14];
            ndims = header[15];
            for ( i = 0; i < MDSDescriptor.MAX_DIM; i ++ )
            {
                dims[i] = byteArrayToInt( header, 16 + 4*i);
            }

            if ( msglen > 48 )
            {
                if ( compressed )
                {
                    body = readCompressedBytes( msglen - 52, s );

                }
                else
                {
                    body = new byte[msglen - 48];
                    s.readFully(body);
```

```
            }

        }
        else
        {
            body = new byte[0];
        }
    }
    /* Convert the message to an MDSDescriptor object.*/
    public MDSDescriptor toDescriptor()
    {
        MDSDescriptor desc;
        int i;
        desc = new MDSDescriptor();
        desc.setDtype(dtype);
        switch ( dtype )
        {
            case MDSDescriptor.DTYPE_CSTRING:
                desc.setCstringData(new String( body ));
            break;
            case MDSDescriptor.DTYPE_CHAR:
                desc.setCharData(new String( body ));
            break;
            case MDSDescriptor.DTYPE_WORDU: //unsigned word
                desc.setIntData(convertBytesToShortToInt());
                desc.setDtype(MDSDescriptor.DTYPE_INT);
            break;
            case MDSDescriptor.DTYPE_BYTE:
                desc.setByteData(body);
            break;
            case MDSDescriptor.DTYPE_SHORT:
                    // will be in error; need conversion
                desc.setIntData(convertBytesToShortToInt());
                desc.setDtype(MDSDescriptor.DTYPE_INT);
            break;
            case MDSDescriptor.DTYPE_INT:
                desc.setIntData(convertBytesToInt());
            break;
            case MDSDescriptor.DTYPE_FLOAT:
                desc.setDoubleData(
                        convertBytesToFloatToDouble());
                desc.setDtype(MDSDescriptor.DTYPE_DOUBLE);
            break;
            case MDSDescriptor.DTYPE_DOUBLE:
                desc.setDoubleData(convertBytesToDouble());
            break;
            case MDSDescriptor.DTYPE_EVENT:
                desc.setEventData(new String( body ));
            break;
```

```java
            default :
                // We'll let the caller handler invalid messages
                desc.setByteData(body);
            break;
        }
        return desc;
    }
    /* Read bytes from a compressed input stream */
    private byte[] readCompressedBytes ( int bytes,
                        DataInputStream s ) throws IOException
    {
        int bytes_to_read;
        int read_bytes;
        int offset;
        byte out[];
        byte b4[];
        InflaterInputStream zs;
        b4 = new byte[4];
        s.readFully(b4);
        bytes_to_read = byteArrayToInt( b4, 0);
        out = new byte[bytes_to_read];
        zs = new InflaterInputStream( s );
        offset = 0;
        while ( bytes_to_read > 0 )
        {
            read_bytes = zs.read( out, offset,
                                bytes_to_read );
            offset += read_bytes;
            bytes_to_read -= read_bytes;
        }
        return out;
    }
    /* Convert message header bytes to ints*/
    private int byteArrayToInt( byte buffer[], int i )
    {
        return
            ( ( buffer[i + 0] & 0xff ) << 24 ) +
            ( ( buffer[i + 1] & 0xff ) << 16 ) +
            ( ( buffer[i + 2] & 0xff ) << 8 ) +
            ( ( buffer[i + 3] & 0xff ) << 0 );

    }
    /* Convert message header bytes to short ints*/
    private short byteArrayToShort( byte buffer[], int i )
    {
        return (short)(( ( buffer[i + 0] & 0xff ) << 8 ) +
            ( ( buffer[i + 1] & 0xff ) << 0 ));
    }
    /* Convert message "body" byte array to an int array */
```

C.4 MDSMessage

```java
private int[] convertBytesToInt()
{
    int i;
    int j;
    int tmp;
    int out[];

    out = new int[body.length / 4];

    for ( i = 0, j = 0; j < body.length; i ++, j += 4 )
    {
        out[i] =
            ( ( body[j + 0] & 0xff ) << 24 ) +
                            //& 0xff converts to int
            ( ( body[j + 1] & 0xff ) << 16 ) +
            ( ( body[j + 2] & 0xff ) << 8 ) +
            ( ( body[j + 3] & 0xff ) << 0 );
    }
    return out;
}
/* Convert message "body" byte array to shorts
   and then to an int array */
private int[] convertBytesToShortToInt()
{
    int i;
    int j;
    int tmp;
    int out[];

    out = new int[body.length / 2];

    for ( i = 0, j = 0; j < body.length; i ++, j += 2 )
    {
            out[i] =
            ( ( body[j + 0] & 0xff ) << 8) +
            ( ( body[j + 1] & 0xff ) << 0);
    }
    return out;
}
/* Convert message "body" byte array to floats
   and then to a double array */
private double[] convertBytesToFloatToDouble()
{
    int i;
    int j;
    int tmp;
    double outDouble[];

    outDouble = new double[body.length / 4];
```

```java
        for ( i = 0, j = 0; j < body.length; i++, j += 4 )
        {
            tmp =
                ( ( body[j + 0] & 0xff ) << 24) +
                ( ( body[j + 1] & 0xff ) << 16) +
                ( ( body[j + 2] & 0xff ) << 8) +
                ( ( body[j + 3] & 0xff ) << 0);
            outDouble[i] = Float.intBitsToFloat( tmp );
        }
        return outDouble;
    }
    /* Convert message "body" byte array to a double array*/
    private double[] convertBytesToDouble()
    {
        long ch;
        double out[] = new double[body.length / 8];
        for(int i = 0, j = 0; i < body.length / 8; i++, j+=8)
        {
            ch  = (body[j+0] & 0xffL) << 56;
            ch += (body[j+1] & 0xffL) << 48;
            ch += (body[j+2] & 0xffL) << 40;
            ch += (body[j+3] & 0xffL) << 32;
            ch += (body[j+4] & 0xffL) << 24;
            ch += (body[j+5] & 0xffL) << 16;
            ch += (body[j+6] & 0xffL) << 8;
            ch += (body[j+7] & 0xffL) << 0;
            out[i] = Double.longBitsToDouble(ch);
        }
        return out;
    }
    public void setNargs(byte _nargs)
    {
        nargs = _nargs;
    }
}
```

D
Listings for PreEScope Examples

D.1 PreEScope0

```java
import java.io.*;
/** This version reads in some MDSplus data and
    writes it to 2 files:
    File "data_yVals":
            Number of y values; expression
    File "data_xVals":
            Number of x values; dim_of(expression)
*/
public class PreEScope0
{
    public static void main(String[] args)
    {
        new PreEScope0(args);
    }
    public PreEScope0(String[] args)
    {
        String server;
        String experiment;
        int shot=0;
        String expression = " ";
        int i,yLen,xLen;
        double[] yVals,xVals;
        MDSNetworkSource dataSource;
        MDSDescriptor result2=null, result4=null;
        /* Make sure all the parameters are given */
        if ( args.length != 4 )
        {
            System.out.println( "Usage: PreEScope0 <server> "
            + "<experiment> <shot> <expression>" );
            System.exit(-1);
        }
```

```java
/* Get the parameters */
server = args[0];
experiment = args[1];
expression = args[3];
try
{
    shot = Integer.parseInt( args[2] );
}
catch ( NumberFormatException e )
{
    System.out.println( "Invalid Shot Number: "
                        + e.getMessage() );
    System.exit(-1);
}
/* Connect to server and open experiment */
dataSource = new MDSNetworkSource();
try
{
    System.out.println(
            "\nConnecting to server ..." );
    dataSource.connect( server );

    System.out.println(
            "Opening Experiment and Shot ..." );
    dataSource.open( experiment, shot );
}
catch ( IOException e )
{
    System.out.println(
        "Failed to communicate with server: " +
    e.getMessage() );
    System.exit(-1);
}
try
{
    System.out.println(
        "Evaluating Expression ..." + expression );
    result2 = dataSource.evaluate( expression );
    if(result2.getDoubleData() == null)
    {
        System.out.println(
        "Failed to retrieve " + expression +
        ": " + result2.getCstringData());
        System.exit(-1);
    }
    expression="dim_of("+expression+")";
    System.out.println(
        "Evaluating Expression ..." + expression );
    result4 = dataSource.evaluate( expression );
```

```java
            if(result4.getDoubleData() == null)
            {
                System.out.println("Failed to retrieve "
                    + expression + ": " +
                    result2.getCstringData());
                System.exit(-1);
            }

    }
    catch ( IOException e )
    {
        System.out.println(
                    "Failed to communicate with server: "
                        + e.getMessage() );
        System.exit(-1);
    }
    //Check for errors when retrieving signals
    try
    {
        File f2 = new File("data_yVals");
        f2.createNewFile();
        DataOutputStream out2 =
            new DataOutputStream(
            new BufferedOutputStream(
            new FileOutputStream(f2)));
        File f3 = new File("data_xVals");
        f3.createNewFile();
        DataOutputStream out3 =
            new DataOutputStream(
            new BufferedOutputStream(
            new FileOutputStream(f3)));
        yLen = result2.getDoubleData().length;
        xLen = result4.getDoubleData().length;
        yVals = new double[yLen];
        xVals = new double[xLen];
        yVals = result2.getDoubleData();
        xVals = result4.getDoubleData();
        out2.writeInt(yLen);
        out3.writeInt(xLen);
        for ( i = 0; i < yLen; i++ )
        {
            out2.writeDouble(yVals[i]);
        }
        for ( i = 0; i < xLen; i++ )
        {
            out3.writeDouble(xVals[i]);
        }
        out2.flush();
        out2.close();
```

```java
            out3.flush();
            out3.close();
            Plotter.plot(xVals,yVals,"test");
        }
        catch (IOException e) {System.out.println(e);}
        /* Close experiment and disconnect from server */
        try
        {
            System.out.println( "\nClosing Experiment ..." );
            dataSource.close();
            System.out.println(
                        "Disconnecting from server ..." );
            dataSource.disconnect();
            dataSource = null;
        }
        catch ( IOException e )
        {
            System.out.println(
                "Failed to communicate with server: "
                + e.getMessage() );
            System.exit(-1);
        }
    }
}
```

D.2 PreEScope1

D.2.1 PreEScope1 Main Program

```java
import javax.swing.*;
public class PreEScope1
{
    public static void main(String args[])
    {
        new PreEScope1();
    }
    public PreEScope1()
    {
        EScopeFrame gui = new EScopeFrame();
        gui.initialise();
    }
}
```

D.2.2 EScopeFrame Class

```java
import javax.swing.*;
import java.awt.event.*;
import java.io.*;
import java.net.*;
import java.util.*;
/** Sets up the main frame for PreEScope1; requests data
from the MDSplus server; calls Plotter to plot data.*/
public class EScopeFrame extends JFrame
{
    private JPanel componentPanel, drawPanel;
    private JMenuItem connectServer;
    private JMenuItem openExperiment;
    private JMenuItem requestData;
    private JMenuItem plotData;
    private JMenuItem quit;
    private PrintWriter out = null;
    private DataInputStream in = null;
    private double[] xVals = null, yVals = null;
    private int xLen = 0, yLen = 0;
    private Socket sock = null;
    private MDSNetworkSource dataSource =
                                    new MDSNetworkSource();
    private String address;
    public EScopeFrame()
    {
        // Components for the GUI
        JMenuBar menuBar = new JMenuBar();
        JMenu menu = new JMenu("Main Menu");
        connectServer =
                    new JMenuItem(new ConnectServerAction());
        openExperiment =
                    new JMenuItem(new OpenExperimentAction());
        requestData = new JMenuItem(new RequestDataAction());
        plotData = new JMenuItem(new PlotDataAction());
        quit = new JMenuItem(new QuitAction());
        menuBar.add(menu);
        menu.add(connectServer);
        menu.add(openExperiment);
        menu.add(requestData);
        menu.add(plotData);
        menu.add(quit);
        setJMenuBar(menuBar);
        connectServer.setEnabled(true);
        openExperiment.setEnabled(true);
        requestData.setEnabled(true);
        plotData.setEnabled(true);
        quit.setEnabled(true);
```

```java
            componentPanel = new JPanel();
    }
    public void initialise()
    {
        getContentPane().add(componentPanel);
        setSize(400,200);
        setLocation(400,400);
        setDefaultCloseOperation(JFrame.EXIT_ON_CLOSE);
        setTitle( "EScope GUI" );
        setVisible(true);
    }
    private class ConnectServerAction extends AbstractAction
    {
        public ConnectServerAction()
        {
            super("Connect Server");
        }
        public void actionPerformed(ActionEvent event)
        {
            String serverString = JOptionPane.showInputDialog
                (EScopeFrame.this,"Specify server and port",
                 "ephebe.anu.edu.au:8000");
            try
            {
                dataSource = new MDSNetworkSource();
                System.out.println(
                            "\nConnecting to server ..." );
                dataSource.connect( serverString );
                System.out.println(
                            "...connection successful");
            }
            catch ( IOException e )
            {
                System.out.println(
                "Failed to communicate with server: " +
                e.getMessage() );
            }
        }
    }
    private class OpenExperimentAction extends AbstractAction
    {
        public OpenExperimentAction()
        {
            super("Open Experiment");
        }
        public void actionPerformed(ActionEvent event)
        {
            int ntokens = 0;
            StringTokenizer strTok=null;
```

```java
            while (ntokens != 2)
            {
                String experimentString =
                            JOptionPane.showInputDialog
                (EScopeFrame.this,
                "Specify experiment and shot",
                "h1data 37025");
                strTok =
                        new StringTokenizer(experimentString);
                ntokens = strTok.countTokens();
            }
            String experiment = strTok.nextToken();
            String portString = strTok.nextToken();
            try
            {
                int shot = Integer.parseInt(portString);
                System.out.println(
                    "Opening Experiment and Shot ..." +
                    experiment + " " + shot);
                dataSource.open( experiment, shot );
                                            // string, int

            }
            catch (NumberFormatException e)
                                {System.out.println(e);}
            catch (IOException e) {System.out.println(e);}
        }
    }
    private class RequestDataAction extends AbstractAction
    {
        public RequestDataAction()
        {
            super("Request Data");
        }
        public void actionPerformed(ActionEvent event)
        {
            address = JOptionPane.showInputDialog
                    (EScopeFrame.this,
                    "Enter legal file address",
                    ".operations:i_fault");
            MDSDescriptor result2=null, result4=null;
            String expression = null;
            try
            {
                expression= address;
                System.out.println( "Evaluating Expression "
                            + expression );
                result2 = dataSource.evaluate( expression );
```

```java
                    expression="dim_of(" + address + ")";
                    System.out.println( "Evaluating Expression  "
                                    + expression );
                    result4 = dataSource.evaluate( expression );
                    yLen = result2.getDoubleData().length;
                    xLen = result4.getDoubleData().length;
                    yVals = new double[yLen];
                    xVals = new double[xLen];

                    yVals = result2.getDoubleData();
                    xVals = result4.getDoubleData();
                }
                catch ( IOException e )
                {
                    System.out.println(
                    "Failed to communicate with server: "
                    + e.getMessage() );
                }
            }
        }
        private class PlotDataAction extends AbstractAction
        {
            public PlotDataAction()
            {
                super("Plot Data");
            }
            public void actionPerformed(ActionEvent event)
            {
                if (xVals != null && yVals != null &&
                    yLen !=0 && xLen !=0 )
                {   Plotter.plot(xVals, yVals, address);}
                else {System.out.println(
                        "Not enough information for plot");}
            }
        }
        private class QuitAction extends AbstractAction
        {
            public QuitAction()
            {
                super("Quit");
            }
            public void actionPerformed(ActionEvent event)
            {
                try
                {
                    if (dataSource.isOpen()) dataSource.close();

                    if (dataSource.isConnected())
                        dataSource.disconnect();
```

```
            System.exit(0);
        }
        catch (IOException e) {System.out.println(e);}
    }
  }
}
```

D.2.3 Plotter

This class is substantially similar to the listing in Appendix B.2 except, as noted in Sec. 3.6, we do not specify `JFrame.EXIT_ON_CLOSE` when the plot frame is created.

D.3 PreEScope2

This is similar to `PreEScope1` with `Plotter` replaced by the following class, `Plotter2`. This is a very substantial, and complicated class. The following listing omits the method `drawYAxis` which is substantially similar to `drawXAxis` apart from rotating the labels and ticks as described in Sec. 4.4. The full source code is available on the CD.

```java
import java.awt.*;
import java.awt.event.*;
import java.awt.geom.*;
import java.awt.font.*;
import javax.swing.*;
import java.text.*;
import java.util.*;
/** Produces a graph with a border, labels, ticks, grid,
    tick−labels and scientific notation. */
public class Plotter2
{
    public static void plot(double[] xVals, double[] yVals,
            String xUnits, String yUnits, String title)
    {   //Note: Do NOT exit application on closing the plot.
        DrawFrame f =
            new DrawFrame(xVals, yVals, xUnits, yUnits, title);
        f.setVisible(true);
    }
}
/* A frame that contains a panel with drawings */
class DrawFrame extends JFrame
{
    public static final int WIDTH = 700;
    public static final int HEIGHT = 500;

    public DrawFrame(double[] xVals, double[] yVals,
            String xUnits, String yUnits, String title)
```

```java
    {
        setTitle(title+"  ("+ yUnits+" versus "+xUnits+")");
        setSize(WIDTH,HEIGHT);
        DrawPanel panel =
            new DrawPanel(xVals,yVals,xUnits,yUnits,title);
        Container contentPane = getContentPane();
        contentPane.add(panel);
    }
}
/* A panel that displays the line */
class DrawPanel extends JPanel
{
    private double[] xVals,yVals;
    private String xUnits,yUnits,title;
    private int borderSize,sigFigs,titleFontHeight,
            axisFontHeight,tickFontHeight,tickPix;
    private double textFac;
    private Font titleFont, axisFont, tickFont;
    private double maxNormalRange = 10.0;
    private double minNormalRange = 1.0;
            // (Do not use sci notation inside normal range)
    private String sciXLabel="";
            //label for x-axis title if sci notation
    private String sciYLabel="";
            //label for y-axis title if sci notati
    /** Nice formater for axis numbers */
    private NumberFormat tickFormatter =
                        NumberFormat.getNumberInstance();

    public DrawPanel(double[] x, double[] y,
        String xu,String yu, String ti)
    {
        this.xVals = x;
        this.yVals = y;
        this.xUnits = xu;
        this.yUnits = yu;
        this.title = ti;
        sigFigs = 5;
        tickFormatter.setMaximumFractionDigits(4);
        tickFormatter.setMinimumIntegerDigits(1);
        titleFontHeight = 16;
        axisFontHeight = 12;
        tickFontHeight = 10;
        textFac = 0.8; // ratio of ascent to height of text
        tickPix = 4; // number of pixels for tick length
        borderSize=0;
            // pixel width of each border to be calculated
            // later
        titleFont = new Font("SansSerif", Font.PLAIN,
```

```java
                                titleFontHeight );
        axisFont = new Font("SansSerif", Font.PLAIN,
                                axisFontHeight );
        tickFont = new Font("SansSerif", Font.PLAIN,
                                tickFontHeight );
    }
    public void paintComponent(Graphics g)
    {
        // Cast the graphics object to Graph2D
        Graphics2D g2 = (Graphics2D) g;
        // Set rendering hints to improve display quality
        g2.setRenderingHint(RenderingHints.KEY_ANTIALIASING,
                    RenderingHints.VALUE_ANTIALIAS_ON );
        // Get plot size
        Dimension size = getSize();
        int xSize = size.width;
        int ySize = size.height;
        // Set background color
        g2.setColor(Color.white);
        g2.fill(
                new Rectangle2D.Double(0,0,size.width,
                                    size.height ));
        // Set the Color and BasicStroke for drawing lines
        g2.setColor(Color.black);
        float strokeWidth = 1.0f;
        float [] solid = {12.0f,0.0f};    // Solid line style
        BasicStroke bs = new BasicStroke( strokeWidth,
                    BasicStroke.CAP_SQUARE,
                    BasicStroke.JOIN_MITER, 1.0f,
                    solid, 0.0f );
        g2.setStroke(bs);
        borderSize = calcBorderSize(g2); // find border size
        /* Perform scaling from data coordinates to pixels */
        double xMax,yMax,xMin,yMin,deltaX,deltaY;
        xMax=xVals[0];
        xMin=xMax;
        yMax=yVals[0];
        yMin=yMax;
        int i=0;
        for (i=1; i<xVals.length; i++)
        {
            if (xVals[i] > xMax) xMax = xVals[i];
            if (yVals[i] > yMax) yMax = yVals[i];
            if (xVals[i] < xMin) xMin = xVals[i];
            if (yVals[i] < yMin) yMin = yVals[i];
        }
        deltaX = xMax − xMin;
        deltaY = yMax − yMin;
        double xScale = (xSize−2∗borderSize)/(deltaX);
```

```java
    double yScale = (ySize-2*borderSize)/(deltaY);
    /* Check how many points might be filtered out
       after transforming to pixels*/
    int npoints = -1;
    int currXVal, currYVal, prevXVal = 0;
    for (i = 0; i < yVals.length; i++)
    { //Find how many points can be filtered
        currXVal = (int) ((xVals[i]-xMin)*xScale
                                    + borderSize);
        if (npoints == -1 || currXVal != prevXVal)
        {
            npoints++;
              //increment counter if x pixel is different
            prevXVal = currXVal;
        }
    }
    // Set up arrays to received scaled points
    int nDim = yVals.length;
    if (npoints <= yVals.length/2) nDim = 2*(npoints +1);
    double[] xScaled = new double[nDim];
    double[] yScaled = new double[nDim];
    if (npoints > yVals.length/2)
    {//Heuristic condition for
     //"not enough points to filter"
        for (i = 0; i < yVals.length; i++)
        {
            xScaled[i] = (int) ((xVals[i]-xMin)*xScale
                                    + borderSize);
            yScaled[i] = (int) (ySize -borderSize
                                    - (yVals[i]-yMin)*yScale);
        }
        npoints = yVals.length;
    }
    else
    { //It makes sense to filter.
      //Adjacent filtered points store max/min Y vals
        npoints = -1;
        for (i = 0; i < yVals.length; i++)
        {
            currXVal = (int) ((xVals[i]-xMin)*xScale
                                    + borderSize);
            currYVal = (int) (ySize - borderSize -
                                    (yVals[i]-yMin)*yScale);
            if (npoints == -1 || currXVal != prevXVal)
            { // Points have different xVals;
              //store two copies
                npoints++;
                prevXVal = currXVal;
                xScaled[2*npoints] = currXVal;
```

```
                        yScaled[2*npoints] = currYVal;
                        xScaled[2*npoints + 1] = currXVal;
                        yScaled[2*npoints + 1] = currYVal;
                    }
                    else
                    { // Points have same xVals;
                      // store max/min yVals
                        if ( currYVal < yScaled[2*npoints])
                                yScaled[2*npoints] = currYVal;
                        if ( currYVal > yScaled[2*npoints+1])
                                yScaled[2*npoints+1] = currYVal;
                    }
                }
                npoints = 2*npoints+1;
            }
            /* Plot curve */
            for ( i=0; i < npoints ; i++)
            {
                Line2D line = new Line2D.Double (xScaled[i],
                        yScaled[i], xScaled[i+1], yScaled[i+1]);
                g2.draw(line);
            }
            /* Draw ticks, grid and tick labels */
            // find height of a dummy axis label
            g2.setFont(axisFont);
            FontRenderContext context =
                                g2.getFontRenderContext();
            Rectangle2D bounds = axisFont.getStringBounds
                                    ("Axis Label", context);
            double axisLabelHeight = bounds.getHeight();
            // change stroke for ticks and grid
            strokeWidth = 0.25f;
            bs = new BasicStroke( strokeWidth,
                        BasicStroke.CAP_SQUARE,
                        BasicStroke.JOIN_MITER, 1.0f,
                        solid, 0.0f );
            g2.setStroke(bs);
            // draw ticks, grid and tick labels and return sci
            // labels
            this.drawXAxis(xSize, ySize, xScale,
                xMin, xMax,(int)axisLabelHeight, g2);
            this.drawYAxis(xSize, ySize, yScale,
                yMin, yMax,(int)axisLabelHeight, g2);
            // draw graph title and axes labels
            drawLabels(titleFont, axisFont, xSize, ySize, g2);

        }
        public int calcBorderSize(Graphics2D g2)
        {
```

```java
        /* borderSize calculation using test string "Test" */
        double border1, border2, border3, spaceFac=1.5;
        // height of top adornment: graph title
        g2.setFont(titleFont); // title font
        FontRenderContext context =
                             g2.getFontRenderContext();
        Rectangle2D bounds = titleFont.getStringBounds(
                                      "Test", context);
        double stringHeight = bounds.getHeight();
        border1= spaceFac*stringHeight;
        // height of bottom adornments: x-axis label,
        // x-tick-labels
        g2.setFont(axisFont); // axis font
        context = g2.getFontRenderContext();
        bounds = axisFont.getStringBounds("Test", context);
        stringHeight = bounds.getHeight();
        border2= spaceFac*stringHeight;
        g2.setFont(tickFont); // tick font
        context = g2.getFontRenderContext();
        bounds = tickFont.getStringBounds("Test", context);
        stringHeight = bounds.getHeight();
        border3= spaceFac*stringHeight;
        return (int) Math.max(border1, border2+border3);
    }
    public void drawLabels(Font titleFont, Font axisFont,
           int xSize, int ySize,
           Graphics2D g2)
    {
        // Graph Title
        g2.setFont(titleFont);
                    // set font and find size of label
        FontRenderContext context =
                             g2.getFontRenderContext();
        Rectangle2D bounds = titleFont.getStringBounds(
                                      title, context);
        double stringWidth = bounds.getWidth();
        double stringHeight = bounds.getHeight();
        int xp = (int)((xSize - stringWidth)/2.0);
        int yp = (int)((borderSize - stringHeight)/2.0 +
                                textFac*stringHeight);
        g2.drawString(title, xp, yp);
        /* Draw X-axis label */
        g2.setFont(axisFont);
                    // set font and find size of label
        context = g2.getFontRenderContext();
        bounds = axisFont.getStringBounds(xUnits+sciXLabel,
                                          context);
        stringWidth = bounds.getWidth();
        stringHeight = bounds.getHeight();
```

```java
        xp = (int)((xSize - stringWidth)/2.0);
        yp = (int)(ySize - 1.25*stringHeight +
                            textFac*stringHeight);
        g2.drawString(xUnits+sciXLabel, xp, yp);
        /* Draw Y-axis label */
        bounds = axisFont.getStringBounds(yUnits+sciYLabel,
                            context);   //font unchanged
        stringWidth = bounds.getWidth();
        stringHeight = bounds.getHeight();
        xp = (int)(stringHeight*0.5+ stringHeight*textFac);
        yp = (int)(ySize - (ySize - stringWidth)/2.0);
        // rotate graphics context; draw the message;
        // rotate back
        g2.rotate(-90.0*Math.PI/180.,xp,yp);
        g2.drawString(yUnits+sciYLabel, xp, yp);
        g2.rotate(+90.0*Math.PI/180.,xp,yp);
}

/** Draw x-axis grid, ticks and tick labels.
    Return sciXLabel */
public void drawXAxis(int xSize, int ySize,
    double xScale, double xMin, double xMax,
    int axisLabelHeight, Graphics2D g2D)
{
    int tickOffset =3;
    // set font and find FontRenderContext for label
    // dimensions
    g2D.setFont(tickFont);
    FontRenderContext context =
                        g2D.getFontRenderContext();
    String dummyTickL = ".";
    // compute width of dummy tick label
    for (int i = 0; i < sigFigs; i++)
    {
        dummyTickL = dummyTickL + "W";
    }
    Rectangle2D bounds = tickFont.getStringBounds(
                        dummyTickL, context);
    int textWidth = (int) bounds.getWidth();

    //find number of ticks
    int numTicks = (int) ((double)(xSize - 2*borderSize)/
                        (double)textWidth);
    if (numTicks<2) numTicks=2;
    int nBin = numTicks-1;   // number of bins

    double dataRange = nicenum(xMax-xMin, true);
    double d = nicenum(dataRange/(nBin), false);
```

```java
// tick range should be smaller than the data range
double tickMin= d*Math.ceil(xMin/d);
double tickMax = d*Math.floor(xMax/d);

// Fill in an arraylist of x tick values
ArrayList xTickPoints = new ArrayList();
for (double x = tickMin; x<= tickMax+0.5*d ; x=x+d)
{
    xTickPoints.add(new Double(x));
}

Double dummyMinTick = (Double) xTickPoints.get(0);
                                //first
double minTick = dummyMinTick.doubleValue();
Double dummyMaxTick = (Double)
            xTickPoints.get(xTickPoints.size()-1);
                                //last
double maxTick = dummyMaxTick.doubleValue();
// Determine whether to use scientific notation;
// find multiplier and sciLabel
double maxAbsTick = Math.max(Math.abs(maxTick),
                    Math.abs(minTick));
int powerOfTen = 0;
double multiplier = 1;
boolean scientific = false;
sciXLabel="";
if ( maxAbsTick >= maxNormalRange ||
     maxAbsTick < minNormalRange)
{
    powerOfTen=(int) Math.floor(
                        Math.log10(maxAbsTick));
    scientific=true;
    sciXLabel = "  (x10^"+powerOfTen+")";
    multiplier = Math.pow(10,powerOfTen);
}
for (int i=0; i<xTickPoints.size(); i++)
{
    Double dummy = (Double) xTickPoints.get(i);
    double dp = dummy.doubleValue();
    // Draw little tick mark
    int gp  = (int)((dp-xMin)*xScale + borderSize);
    int yu = (int) ySize-borderSize;
    int yl = yu+tickPix;
        // Number of pixels below the axis for tick
    g2D.drawLine(gp,yu,gp,yl); // Draw little tick
    // Draw grid line
    g2D.drawLine(gp,yu,gp,borderSize);
    // Find tick label
    String tickLabel="";
```

```java
            double tickNumber = 0.;
            if (scientific)
            {   tickNumber = dp/multiplier;}
                    // Divide by exponent to get number
            else
            {   tickNumber = dp;}
            tickLabel = tickFormatter.format(tickNumber);

            // Calc tick label width and height
            bounds = tickFont.getStringBounds(tickLabel,
                                                context);
            double stringWidth = bounds.getWidth();
            double stringHeight = bounds.getHeight();
            // Draw label
            int xp = gp - (int) (stringWidth/2.0);
            int yp = (int) (ySize - borderSize + tickOffset
                                + textFac*stringHeight);
            g2D.drawString(tickLabel,xp,yp);
        }
    }


// drawYAxis listing omitted........

/** Converts a given number into a 'nice number'
    @param x The number to convert
    @param round True for rounded to closest;
     false for rounded up
    @return The nice number near x */
private double nicenum(double x, boolean round)
{
    int exp;
    double f,nf;
    if (x <= 0.)
    {
        System.out.println(
                "Illegal value passed to nicenum: " + x);
        System.exit(0);
    }
    exp = (int) Math.floor(Math.log10(x));
    f = x/Math.pow(10.0,exp);
    if (round) // round to closest nice number
    {
        if (f<1.5){ nf=1.0;}
        else
        {   if (f<3.0)
            {   nf=2.0;
            }
            else
```

```
            {     if (f<7.0)
                  {    nf=5.0;
                  }
                  else
                  {    nf=10.0;
                  }
            }
        }
    }
    else   // round up to nice number
    {     if (f<=1.0){ nf=1.0;}
          else
          {    if (f<=2.0)
               {    nf=2.0;
               }
               else
               {    if (f<=5.0)
                    {    nf=5.0;
                    }
                    else
                    {    nf=10.0;
                    }
               }
          }
    }
    return  nf*Math.pow(10.0,exp);
}
```

D.4 MDSTree and MDSTreeNode

The following listings are relevant to all versions of `PreEScope` and `EScope` from `PreEScope4` onwards.

D.4.1 `MDSTreeNode`

```
/** Extends DefaultMutableTreeNode to store the node type and
    overrides toString for node rendering
*/
public class MDSTreeNode extends DefaultMutableTreeNode
{
    public static final int TYPE_TOP    = 0;
    public static final int TYPE_MEMBER = 1;
    public static final int TYPE_CHILD  = 2;
    private String name;
    private int type;
```

```java
/**
Creates a new tree node with no child or member nodes
@param _name The name of this node
@param _type The type of this node (top, member or child)
*/
public MDSTreeNode(String _name, int _type)
{
    super();
    name = _name;
    type = _type;
}
public MDSTreeNode()
{
    super();
}
public String getName()
{
    return name;
}
public int getType()
{
    return type;
}
/**
The name of this node,
with a ':' if node is a member,
or a '.' if node is a child
@return A text representation of this node
*/
public String toString()
{
    if ( type == TYPE_MEMBER )
    {
        return ":" + name;

    }
    else if ( type == TYPE_CHILD )
    {
        return "." + name;

    }
    else
    { //top node
        return name;
    }
}
}
```

D.4.2 MDSTree

```java
import java.io.*;
import java.util.*;
import javax.swing.tree.DefaultTreeModel;

/** Connects to MDSplus database and
    constructs the experiment tree */
public class MDSTree extends DefaultTreeModel
{
    private MDSDataSource database;
    public MDSTree(MDSDataSource _database,
                MDSTreeNode _top)
    {
        super(_top);
        database = _database;
    }
    /** Recursively create an experiment tree
        from the MDSPlus server
        @throws IOException If not connected
        to server or experiment not opened */
    public void getTree() throws IOException
    {
        String experiment;
        if ( database != null && database.isConnected()
            && database.isOpen() )
        {
            experiment = database.getExperiment();

            MDSTreeNode top = getSubTree("\\\\" +
                    experiment.toUpperCase() + "::TOP",
                    "TOP", MDSTreeNode.TYPE_TOP );
            setRoot(top);
        }
        else
        {
            throw new IOException( "Database not open" );
        }
    }
    /** Part of the recursion algorithim used to create the
        tree
        @param path The path to the parent node of the sub-tree
        @param nodeName The name of this node
        @param type The node type
        @return The newly-created parent node of this sub-tree
        @throws IOException From call to getSubTreeList
    */
    private MDSTreeNode getSubTree(String path,
            String nodeName, int type) throws IOException
```

```
{
    MDSTreeNode out;
    String nodelist[];
    int i;
    out = new MDSTreeNode( nodeName, type );
    // Process MDSplus "member" nodes
    nodelist = getSubTreeList( path + ":*" );
    if ( nodelist != null )
    {
        for ( i = 0; i < nodelist.length; i++ )
        {
            out.add(getSubTree(path + ":" + nodelist[i],
                nodelist[i], MDSTreeNode.TYPE_MEMBER ) );
        }
    }
    // Process MDSplus "child" nodes
    nodelist = getSubTreeList( path + ".*" );
    if ( nodelist != null )
    {
        for ( i = 0; i < nodelist.length; i++ )
        {
            out.add(getSubTree(path + "." + nodelist[i],
                nodelist[i], MDSTreeNode.TYPE_CHILD ) );
        }
    }
    //System.out.println(out);
    //Uncomment to print node names.
    return out;
}
/**
Gets a list of all member and child nodes of a node
@param path The path to the parent node
@return A list of the member and child nodes
@throws IOException If there is an error
communicating with the MDSPlus server
*/
private String[] getSubTreeList(String path)
                                    throws IOException
{
    MDSDescriptor result;
    int nodes;
    int i;
    String out[];
    String expr;
    expr = "GETNCI(\"" + path + "\",\"NODE_NAME\")";
    result = database.evaluate( expr );
    if ( result.getDtype() ==
                        MDSDescriptor.DTYPE_CSTRING )
    {
```

```
            if (
              result.getCstringData().endsWith(
                                         "Node Not Found"))
            { //There are no nodes below this path
                return null;
            }
            if ( result.getCstringData().substring( 0, 4 )
                                            .equals( "Tree" ))
            { //Some other sort of error: throw exception
                throw new IOException( "GETNCI Error: " +
                                    result.getCstringData() );
            }

            nodes = result.getCstringData().length() / 12;
            out = new String[nodes];
            for ( i = 0; i < nodes; i ++ )
            {
                out[i] =
                    result.getCstringData().substring( 12*i,
                    12*(i+1) ).trim();
            }
        }
        else
        { // Didn't get correct format from message
            throw new IOException( "GETNCI Error: " +
                            "Incorrect format returned: " +
                            result.getDtype() );
        }
        return out;
    }
}
```

D.5 PreEScope4

Compared with `PreEScope3`, this version of the software has the following modifications:

- `MDSNetworkSource` has a small amount of additional coding to construct a Swing `JTree` and to return this object to the GUI.
- The `MDSDataSource` interface has to be changed to include the signatures of two methods, `getJTree` and `constructTree`.
- The `MDSTree` and `MDSTreeNode` classes, of the previous section, need to be included.
- The `Plotter3` class becomes the `GraphPanel` class. In all honesty, the substantial differences between these two classes is not very great. One is a stand-alone window and the other is a panel. As discussed in Sec. 6.6, we also have used this revision to improve the plotting to deal with possible errors in the data being

sent from MDSplus. So the GraphPanel class is somewhat more complicated than Plotter3.
- There is more work to do by the GUI to tie the whole application together. So the EScopeFrame class is more complicated.
- As described in Sec. 6.6.1, the main GUI pops up stand-alone dialog boxes to request the server address and experiment details. These classes are ConnectDialog and OpenDialog.

D.5.1 EScopeFrame

```java
import java.io.*;
import java.awt.*;
import java.awt.event.*;
import javax.swing.*;
import javax.swing.event.*;
import javax.swing.tree.*;
import java.awt.geom.*;
import java.awt.font.*;
import java.text.*;
import java.util.*;
public class EScopeFrame extends JFrame
{
    private MDSDataSource database;
    private String experiment; // experiment name
    private String shortFileAddress;
                                // for retrieving plot data
    private double[] xVals=null, yVals=null;
                                // plot data arrays
    private String xUnits=null, yUnits=null;
    private int xLen=0, yLen=0;
    private boolean isGraphData=false;
                                //Is there data to plot?
    private boolean isDataError=false;
                                //Flags unexpected data from MDSPlus
    // Components for the menu
    private JMenuBar menuBar;
    private JMenu fileMenu;
    private JMenuItem fileConnectItem;
    private JMenuItem fileOpenItem;
    private JMenuItem fileExitItem;
    // Components for the main window
    private GraphPanel graph;
    private JLabel description;// file address
    private JSplitPane rightPane;
    private JSplitPane mainPane;
    // Dialogs
    private OpenDialog openD;
    private ConnectDialog connectD;
```

```java
// file tree
private JTree tree;
public EScopeFrame()
{
    // Construct the menu
    menuBar = new JMenuBar();
    setJMenuBar( menuBar );
    fileMenu = new JMenu( "File" );
    menuBar.add( fileMenu );
    fileConnectItem = new JMenuItem("Connect to server");
    fileConnectItem.addActionListener(
                              new ConnectAction());
    fileMenu.add( fileConnectItem );
    fileOpenItem = new JMenuItem("Open experiment");
    fileOpenItem.addActionListener(new OpenAction());
    fileMenu.add( fileOpenItem );
    fileMenu.addSeparator();
    fileExitItem = new JMenuItem("Exit");
    fileExitItem.addActionListener(new ExitAction());
    fileMenu.add( fileExitItem );
    // Construct the main pane
    JPanel descriptionPanel = new JPanel();
    descriptionPanel.setLayout(new GridLayout(1,3));
    description = new JLabel();
    descriptionPanel.add(description);
    graph = new GraphPanel(this);
    rightPane = new JSplitPane(
                JSplitPane.VERTICAL_SPLIT,
                new JScrollPane( description ),graph);
    connectD = new ConnectDialog(this,
                    "Enter Connection Details", true);
    openD = new OpenDialog(this,
                    "Enter experiment and shot", true);
}
public void initialise(MDSDataSource _source)
{
    database = _source;
    tree=database.getJTree();
    tree.addTreeSelectionListener(
                    new DataTreeSelectionListener() );
    JScrollPane treeScroll = new JScrollPane( tree );
    treeScroll.setMinimumSize( new Dimension( 150, 10 ) );
    mainPane = new JSplitPane(
                JSplitPane.HORIZONTAL_SPLIT,
                treeScroll, rightPane );
    getContentPane().add( mainPane, "Center" );
    this.setDefaultCloseOperation(JFrame.EXIT_ON_CLOSE);
    this.setTitle( "No experiment opened" );
    this.setBounds( 200, 200, 800, 400 );
```

```java
        this.show();
    }
    public void graphPointUpdated(double x, double y,
                                        boolean inRange)
    {
        if (inRange)
        {
            DecimalFormat decForm =
                            new DecimalFormat("0.##E0");
            String xString = decForm.format(x);
            String yString = decForm.format(y);
            description.setText(" x= " + xString + "   y= "
                                                + yString);
        }
        else description.setText(" ");
    }
    public void setDescription(String text)
    {
        description.setText(text);
    }
    public void getPlotData()
    {
        MDSDescriptor result1=null, result2=null,
            result3=null, result4=null;
        String expression1="", expression2="",
            expression3="", expression4="";
        try
        {
            expression1=" units_of(" +
                        shortFileAddress + ")"; // Y title
            result1 = database.evaluate( expression1 );
            expression2= shortFileAddress;  //Y values
            result2 = database.evaluate( expression2 );
            expression3=" units_of(dim_of(" +
                        shortFileAddress +"))"; // X title
            result3 = database.evaluate( expression3 );
            expression4="dim_of(" +
                        shortFileAddress + ")"; // X values
            result4 = database.evaluate( expression4 );
        }
        catch ( IOException e )
        {
            System.out.println(
                    "Failed to communicate with server: "
                    + e.getMessage() );
            System.exit(-1);
        }
        // check x and y axis label types
        if ((result1.getDtype()!=result1.DTYPE_CSTRING) ||
```

```java
            (result3.getDtype()!=result3.DTYPE_CSTRING))
        {
            System.out.println("Error:
                String descriptor not returned for one of:");
            System.out.println("            " +
                expression1 + " and " + expression3);
            System.out.println("(should never happen if
                            MDSPlus database is correct)");
            isDataError=true;
        }
        // get x and y labels
        yUnits = result1.getCstringData();
        xUnits = result3.getCstringData();
        // check xVals and yVals types
        // - double => legal data
        // - strings => some data does not exist;
        // No y data => no data
        if ((result2.getDtype()!=result2.DTYPE_CSTRING) &&
            (result2.getDtype()!=result2.DTYPE_DOUBLE))
        {
            System.out.println("Error: Double or String
                            descriptor not returned for:");
            System.out.println("  yVals:         " +
                            expression2);
            System.out.println("Descriptor type: " +
                result2.getDtype());
            isDataError=true;
        }
        if ((result4.getDtype()!=result4.DTYPE_CSTRING) &&
            (result4.getDtype()!=result4.DTYPE_DOUBLE))
        {
            System.out.println("Error: Double or String
                            descriptor not returned for:");
            System.out.println("  xVals:         " +
                            expression4);
            System.out.println("Descriptor type: " +
                            result4.getDtype());
            isDataError=true;
        }
        if ((!isDataError) && (result2.getDtype()==
                            result2.DTYPE_DOUBLE))
        { // legal y data exists
            isGraphData = true;
            yLen = result2.getDoubleData().length;
            boolean isXData = false;
            if (result4.getDtype()==result4.DTYPE_DOUBLE)
            {
                isXData = true;
                xLen = result4.getDoubleData().length;
```

```
        }
        else
        {
            xLen = yLen;
        }
        if (xLen!=yLen)
        {
            System.out.println("Warning:
                            xLen not yLen in data");
            System.out.println("xLen = " + xLen +
                            " yLen=" + yLen);
            System.out.println(
                        "(xData will be ignored for plot)");
            xLen = yLen;
            isXData = false;
        }
        yVals = new double[yLen];
        xVals = new double[yLen];
        for (int i = 0; i < yLen; i++)
        {
            yVals[i] = result2.getDoubleDataElement(i);
        }
        for (int i = 0; i < yLen; i++)
        {
            if (isXData)
            {
                xVals[i] =
                        result4.getDoubleDataElement(i);
            }
            else
            {
                xVals[i] = i;
            }
        }
    }
}
private boolean hasSignal(String fullPath)
{
    String path1 =
        "GETNCI(\"\\"+fullPath+"\",\"USAGE_SIGNAL\")";
    boolean ok = false;
    try
    {
        MDSDescriptor result = database.evaluate(path1);
        if (result.getDtype()==result.DTYPE_BYTE)
        {
            if (result.getByteData()[0]==1)
            {
                ok = true;
```

```java
                }
            }
        }
        catch( Exception e )
        {
            System.out.println(
                "cannot find if node has a signal : "+
                e.getMessage());
        }
        return ok;
    }
    // Private Inner Classes
    private class ExitAction implements ActionListener
    {
        public ExitAction() {}
        public void actionPerformed( ActionEvent event )
        {   System.exit( 0 );
        }
    }
    private class ConnectAction implements ActionListener
    {
        public ConnectAction(){}
        public void actionPerformed( ActionEvent event )
        {
            if (connectD.showDialog())
            {
                String server = connectD.getServer();
                try
                {
                    if ( database.isConnected() )
                    {
                        database.disconnect();
                    }
                    database.connect( server );
                    String newTitle =
                        "Connection made.
                        No experiment opened";
                    EScopeFrame.this.setTitle(newTitle);
                }
                catch (IOException e)
                {
                    JOptionPane.showMessageDialog(
                        EScopeFrame.this,
                        "Failed to connect to server",
                        "Error",
                        JOptionPane.ERROR_MESSAGE );
                }
            }
        }
```

```java
}
private class OpenAction implements ActionListener
{
    public OpenAction(){}
    public void actionPerformed( ActionEvent event )
    {   if (openD.showDialog())
        {
            experiment = openD.getExperiment();
            int shot = openD.getShot();
            try
            {
                if ( database.isOpen() )
                {
                    database.close();
                }
                database.open( experiment, shot );
                database.constructTree();
                String newTitle =
                    " Opened experiment: "
                    + experiment + " shot: " + shot;
                EScopeFrame.this.setTitle(newTitle);
            }
            catch ( IOException e )
            {
                JOptionPane.showMessageDialog(
                  EScopeFrame.this,
                  "Failed to open " +
                  "experiment and shot:" +
                  e.getMessage(),
                  "Error", JOptionPane.ERROR_MESSAGE );
            }
        }
    }
}
/**    Plots the graph and displays the
       path of the selected tree node */
private class DataTreeSelectionListener
                    implements TreeSelectionListener
{
    public void valueChanged( TreeSelectionEvent e )
    {   // Get new data and plot it
        Object path[];
        String stringPath="";
        path = e.getPath().getPath();
        shortFileAddress="";
        for ( int i = 1; i < path.length; i++ )
        {
            shortFileAddress = shortFileAddress +
                                    path[i].toString();
```

```
                }
                stringPath = stringPath + shortFileAddress;
                if (!shortFileAddress.equals(""))
                    // (Do not plot or display the "TOP" node)
                {
                    EScopeFrame.this.setDescription(stringPath);
                    isGraphData = false;
                    isDataError = false;
                    if (EScopeFrame.this.hasSignal(stringPath))
                        EScopeFrame.this.getPlotData();
                    EScopeFrame.this.graph.setGraphData
                        (xVals, yVals, xUnits, yUnits,
                         shortFileAddress, isGraphData, isDataError);
                    EScopeFrame.this.graph.repaint();
                            //(prints a message if no data)
                }
            }
        }
    }
}
```

D.5.2 ConnectDialog

```
import java.awt.*;
import java.awt.event.*;
import javax.swing.*;
public class ConnectDialog extends JDialog
{
    private JTextField serverField;
    private JTextField portField;
    private JButton okButton;
    private JButton cancelButton;
    private String server;
    private int port;
    private boolean okPressed;
    /** @param mode Is this dialog modal? */
    public ConnectDialog(JFrame parent, String title,
                    boolean mode)
    {
        super( parent, title, mode );
        Container mainPane = getContentPane();
        mainPane.setLayout( new BorderLayout() );
        // data panel
        JPanel dataPanel = new JPanel();
        dataPanel.setLayout( new GridLayout( 2, 2 ) );
        serverField = new JTextField( "localhost" );
        portField = new JTextField( "8000" );
        dataPanel.add( new JLabel( "Server" ) );
```

```java
        dataPanel.add( serverField );
        dataPanel.add( new JLabel( "Port" ) );
        dataPanel.add( portField );
        mainPane.add( dataPanel, BorderLayout.CENTER );
        // button panel
        JPanel buttonPanel = new JPanel();
        okButton = new JButton( "Ok" );
        okButton.addActionListener( new ActionListener()
        {
            public void actionPerformed( ActionEvent event )
            {
                if ( validateFields() )
                {
                    okPressed = true;
                    setVisible( false );
                }
            }


        }
        );
        cancelButton = new JButton( "Cancel" );
        cancelButton.addActionListener( new ActionListener()
        {
            public void actionPerformed( ActionEvent event )
            {
                setVisible( false );
            }


        }
        );
        buttonPanel.add( okButton );
        buttonPanel.add( cancelButton );
        mainPane.add( buttonPanel, BorderLayout.SOUTH );
        setBounds( 200, 200, 300, 200 );
    }

    /** Make the dialog visible and set okPressed to
        "false"*/
    public boolean showDialog()
    {
        okPressed=false;
        this.show();
        return okPressed;
    }
    /** Checks the server and port fields for valid input
        @return "true" if fields are valid,
        "false" otherwise */
    private boolean validateFields()
    {
```

```java
            String porttmp;
            try
            {
                server = serverField.getText();
                porttmp = portField.getText();

            }
            catch ( NullPointerException e )
            {
                JOptionPane.showMessageDialog( this,
                    "Error retrieving data from fields",
                    "Invalid Entry",
                    JOptionPane.ERROR_MESSAGE );
                return false;
            }
            try
            {
                port = Integer.parseInt( porttmp );
            }
            catch ( NumberFormatException e )
            {
                JOptionPane.showMessageDialog( this,
                    "Invalid Port Number",
                    "Invalid Entry",
                    JOptionPane.ERROR_MESSAGE );
                return false;
            }
            if ( port <= 0 || port >= 65536 )
            {
                JOptionPane.showMessageDialog( this,
                    "Invalid Port Number",
                    "Invalid Entry",
                    JOptionPane.ERROR_MESSAGE );
                return false;
            }
            return true;
        }
        // get and set methods
        /** @return The server and port separated by a colon */
        public String getServer()
        {
            return server + ":" + port;
        }
}
```

E
Listing for EScope4

E.1 Package Structure

There is a two-level package structure to this program. The top-level directory contains the main program, EScope4, which is shown below. There are three subdirectories which contain concrete classes:

- dataServerDomain
- graphicsDomain
- guiDomain

  and there are two subdirectories which contain interfaces and abstract classes:

- sharedDataInterfaces
- sharedInterfaces

  The main program reads:

```
import sharedInterfaces.*;
import dataServerDomain.*;
import guiDomain.*;
import graphicsDomain.*;
import javax.swing.*;
public class EScope4
{
    public static void main(String args[])
    {
        new EScope4();
    }
    public EScope4()
    {   // create domains
        GuiFacadeInterface gui = new GuiFacade();
        AbstractGraphicsFacade graph = new GraphicsFacade();
        DataServerFacadeInterface data =
                                new DataServerFacade();
        // associate domains with each other
```

```
            gui.initialise(data, graph);
            graph.initialise(gui);
        }
}
```

E.2 Shared Data Interfaces

There are two interfaces in this package

- GraphDataInterface
- GraphOptionsInterface

The GraphDataInterface reads:

```
package sharedDataInterfaces;
public interface GraphDataInterface
{   // utility methods
    /** Return length of y data array*/
    public int getLength();
    /** Returns the smallest x axis data point*/
    public double getMinX();
    /** Returns the largest x axis data point*/
    public double getMaxX();
    /** Returns the smallest y axis data point*/
    public double getMinY();
    /** Returns the largest y axis data point*/
    public double getMaxY();

    // other get methods
    public double[] getX();   // X data
    public double[] getY();   // Y data
    public String getXLabel();
    public String getYLabel();
    public String getTitle();
    public boolean getIsGraphData();
    public boolean getIsDataError();
}
```

The GraphOptionsInterface reads:

```
package sharedDataInterfaces;

import javax.swing.*;
import java.awt.*;
import java.awt.event.*;
import java.util.*;
import javax.swing.border.*;

/** Stores the options settings for the graph panel */
```

```java
public interface GraphOptionsInterface
{
    public static final int numColors = 13;
    public static final int BLACK = 0;
    public static final int WHITE = 1;
    public static final int RED = 2;
    public static final int BLUE = 3;
    public static final int ORANGE = 4;
    public static final int GRAY = 5;
    public static final int LGRAY = 6;
    public static final int DGRAY = 7;
    public static final int MAGENTA = 8;
    public static final int GREEN = 9;
    public static final int YELLOW = 10;
    public static final int CYAN = 11;
    public static final int PINK = 12;

    public static final  String[] colorNames =
        {"Black","White","Red","Blue"
        ,"Orange","Gray","L. Gray","D. Gray",
        "Magenta","Green","Yellow",
                                    "Cyan","Pink"};

    //Display mode constants and variable
    public static final int ZOOM_MODE = 1,
                    CROSSHAIR_MODE = 2, GRAB_MODE = 3;


    /** returns a Color object corresponding to index i */
    public Color selectColor(int i);
    /** returns a string colour name
                            corresponding to index i */
    public String selectColorName(int i);

    public int getTitleColor();

    public int getGraphColor();

    public int getLineColor();

    public boolean getDefaultBorder();

    public boolean getNiceNumbers();

    public int getDisplayMode();
}
```

E.3 Shared Interfaces

There are three interfaces in this package:

- DataServerFacadeInterface
- GraphicsFacadeInterface
- GuiFacadeInterface

and one abstract class:

- AbstractGraphicsFacade

E.3.1 DataServerFacadeInterface

The DataServerFacadeInterface reads:

```
package sharedInterfaces;
import sharedDataInterfaces.GraphDataInterface;

import java.io.*;
import javax.swing.JTree;

    public void connect(String serverAddrCPort )
                                         throws IOException;

    public void disconnect() throws IOException;

    public void open(String experiment, int shot)
                                         throws IOException;

    public void close() throws IOException;

    public void constructTree(JTree tree) throws IOException;

    // get and set and get-status methods
    public boolean isConnected();

    public boolean isOpen();

    public String getExperiment();

    public int getShot();

    public GraphDataInterface getPlotData(String path);

}
```

E.3.2 GraphicsFacadeInterface

The `GraphicsFacadeInterface` reads:

```java
package sharedInterfaces;
import sharedDataInterfaces.*;

public interface GraphicsFacadeInterface
{   /** Plots arrays of data against each other
        @isGraphData when "false" a message is
          output rather than a graph
        @isDataError when "true" an appropriate
          message is written*/
    public void initialise(GuiFacadeInterface guiFacade);
    public void setGraphData(GraphDataInterface data);

    public void setGraphOptions(GraphOptionsInterface gO);
    public void applyGraphOptions();
}
```

E.3.3 GuiFacadeInterface

The `GuiFacadeInterface` reads:

```java
package sharedInterfaces;

public interface GuiFacadeInterface
{
    /** Associates with DataServer domain and completes
        initialisation of the Gui. Shows the GUI */
    public void initialise(DataServerFacadeInterface data,
                           AbstractGraphicsFacade graph);
    public void graphPointUpdated(double x,
                           double y, boolean inRange);
}
```

E.3.4 AbstractGraphicsFacade

The `AbstractGraphicsFacade` class reads:

```java
package sharedInterfaces;

import javax.swing.*;

public abstract class AbstractGraphicsFacade extends JPanel
                  implements GraphicsFacadeInterface
{
}
```

E.4 The Data Server Package

This package has six classes:

- DataServerFacade
- GraphData
- MDSDescriptor
- MDSMessage
- MDSTree
- MDSTreeNode

E.4.1 DataServerFacade

Part of the DataServerFacade class is listed here. See also the listing for MDSNetworkSource in Appendix C.3.

```
package dataServerDomain;

import sharedInterfaces.*;
import sharedDataInterfaces.*;

import java.io.*;
import java.net.*;
import java.util.*;
import javax.swing.*;

public class DataServerFacade
    implements DataServerFacadeInterface
{
    private String serverAddrCPort;
    private String experiment;
    private int shot;

    private boolean isConnected;
    private boolean isOpen;

    private Socket socket;
                // socket for connecting to MDSPlus server
    private DataOutputStream output;
                // output stream for socket
    private DataInputStream input; // input stream for socket

    private MDSTree treeModel = null;
                // tree model to return to GUI
    private MDSTreeNode treeNode;

    public DataServerFacade()
    {
        serverAddrCPort = null;
```

```java
        experiment = null;
        shot = -1;

        isConnected = false;
        isOpen = false;

        socket = null;
        input = null;
        output = null;
        treeNode = new MDSTreeNode();
        treeModel = new MDSTree(this, treeNode);
    }
    public void constructTree(JTree tree) throws IOException
    {
        treeModel.getTree();
        tree.setModel(treeModel);
    }
    ...
    //methods connect, open, disconnect, close, evaluate,
    //sendMessage as for MDSNetworkSource
    ...
    public GraphDataInterface getPlotData(
                        String shortFileAddress)
    {
        MDSDescriptor result1 = null, result2 = null,
                    result3 = null, result4 = null;
        String expression1 = "", expression2 = "",
            expression3 = "", expression4 = "";
        boolean hasGraphData = false, isDataError = false;
        double[] xVals = null, yVals = null;
                                    // plot data arrays
        String xUnits = null, yUnits = null, title = null;
        int xLen = 0, yLen = 0;
        String fullPath = "\\" +
            getExperiment().toUpperCase() + "::" + "TOP";
        fullPath = fullPath + shortFileAddress;
        // try to read the data
        try
        {
            expression1 = "units_of(" +
                    shortFileAddress + ")"; // Y title
            result1 = evaluate(expression1);
            expression2 = shortFileAddress; //Y values
            result2 = evaluate(expression2);
            expression3 = "units_of(dim_of(" +
                    shortFileAddress + "))"; //X title
            result3 = evaluate(expression3);
            expression4 = "dim_of(" +
                    shortFileAddress + ")"; // X values
```

```java
            result4 = evaluate(expression4);
        }
        catch (IOException e)
        {
            System.out.println(
                    "Failed to communicate with server:"
                    +" " + e.getMessage());
            System.exit(-1);
        }
        // check x and y axis label types
        if ( (result1.getDtype() != result1.DTYPE_CSTRING) ||
            (result3.getDtype() != result3.DTYPE_CSTRING))
        {
            System.out.println("Error: "
                + "String descriptor not returned for one of:");
            System.out.println("              " +
                expression1 + " and " + expression3);
            System.out.println(
                "(should never happen if"
                +" MDSPlus database is correct)");
            isDataError = true;
        }
        // get x and y labels
        yUnits = result1.getCstringData();
        xUnits = result3.getCstringData();
        // check xVals and yVals types
        // - double => legal data
        // - strings => some data does not exist;
        // No y data => no data
        if ( (result2.getDtype() != result2.DTYPE_DOUBLE))
        {
            System.out.println("Error: "
                + "Double descriptor not returned for:");
            System.out.println(" yVals:           " +
                                                expression2);
            System.out.println("Descriptor type: " +
                                            result2.getDtype());
            isDataError = true;
        }
        if ( (result4.getDtype() != result4.DTYPE_CSTRING) &&
            (result4.getDtype() != result4.DTYPE_DOUBLE))
        {
            System.out.println(
                    "Error: Double or String " +
                    "descriptor not returned for:");
            System.out.println(" xVals:           " +
                                                expression4);
            System.out.println("Descriptor type: " +
                                            result4.getDtype());
```

```java
            isDataError = true;
    }
    if ( (!isDataError) && (result2.getDtype()
                                == result2.DTYPE_DOUBLE))
    { // legal y data exists
        yLen = result2.getDoubleData().length;
        boolean isXData = false;
        if (result4.getDtype() == result4.DTYPE_DOUBLE)
        {
            isXData = true;
            xLen = result4.getDoubleData().length;
        }
        else
        {
            xLen = yLen;
        }
        if (xLen != yLen)
        {
            System.out.println("Warning: xLen not"
                        + " equal to yLen in data");
            System.out.println("xLen = " + xLen
                                + " yLen= " + yLen);
            System.out.println(
                    "(xData will be ignored for plot)");
            xLen = yLen;
            isXData = false;
        }
        yVals = new double[yLen];
        xVals = new double[yLen];
        for (int i = 0; i < yLen; i++)
        { // get y data
            yVals[i] = result2.getDoubleDataElement(i);
        }
        for (int i = 0; i < yLen; i++)
        {
            if (isXData) // get x data
            {
                xVals[i] =
                        result4.getDoubleDataElement(i);
            }
            else // set x array to indices
            {
                xVals[i] = i;
            }
        }
    }
    title = shortFileAddress;
    // create a new GraphData object
    GraphData data = new GraphData();
```

```
            data.initialise(xVals, yVals, xUnits, yUnits,
                    shortFileAddress, getShot(), isDataError);
            return data;
        }
    }
```

E.4.2 GraphData

The GraphData class reads:

```
package dataServerDomain;
import sharedDataInterfaces.GraphDataInterface;

/** Holds plot data, labels, title and shot number*/
public class GraphData implements GraphDataInterface
{
    private double[] x={},y={};
    private String xLabel=" ",yLabel=" ",title=" ";
    private int shotNo=0;
    private boolean isGraphData=false, isDataError=false;

    public GraphData(){}
    public void initialise(double[] xv, double[] yv,
                    String xt, String yt,
                    String tit, int shot, boolean dataError)
    {
            x = xv;
            y = yv;
            xLabel = xt;
            yLabel = yt;
            title=tit;
            shotNo=shot;
            isGraphData=true;
            isDataError=dataError;
            if (isDataError) isGraphData=false;
    }
    /** Return length of y data array*/
    public int getLength()
    {
        return y.length;
    }
    /** Returns the smallest x axis data point*/
    public double getMinX()
    {
        double minX = x[0];
        for (int i=1; i<x.length; i++)
        {
            if (x[i]<minX) minX=x[i];
```

```java
    }
    return minX;
}
/** Returns the largest x axis data point*/
public double getMaxX()
{
    double maxX = x[0];
    for (int i=1; i<x.length; i++)
    {
        if (x[i]>maxX) maxX=x[i];
    }
    return maxX;
}
/** Returns the smallest y axis data point*/
public double getMinY()
{
    double minY = y[0];
    for (int i=1; i<y.length; i++)
    {
        if (y[i]<minY) minY=y[i];
    }
    return minY;
}
/** Returns the largest y axis data point*/
public double getMaxY()
{
    double maxY = y[0];
    for (int i=1; i<y.length; i++)
    {
        if (y[i]>maxY) maxY=y[i];
    }
    return maxY;
}
// other get methods
public double[] getX(){ return x;}
public double[] getY(){ return y;}
public String getXLabel(){ return xLabel;}
public String getYLabel(){ return yLabel;}
public String getTitle(){ return title;}
public boolean getIsGraphData(){ return isGraphData;}
public boolean getIsDataError(){ return isDataError;}
}
```

E.5 The Graphics Domain

There are 13 classes in this package. These are made up of the decorator classes:

- `GraphDecorator`

- DrawAxisTicks
- DrawAxisTicksX
- DrawAxisTicksY
- DrawCentredMessage
- DrawCrossHair
- DrawLabels
- DrawWaveform
- DrawZoomBox

and the two adapter classes:

- GraphDataInGraphics
- GraphOptionsInGraphics

and the two substantive classes:

- GraphMediator
- GraphicsFacade

E.5.1 The Decorator Classes

The abstract superclass for the decorator chain reads:

```
package graphicsDomain;
import java.awt.*;
//Implementation of the decorator pattern
public abstract class GraphDecorator
{
    private GraphDecorator decorator=null;
                            //decoration chain component
    protected GraphMediator med;
    public GraphDecorator(GraphMediator m) { this.med =m;}
    public void setDecorator(GraphDecorator decorator)
    {
        this.decorator = decorator;
    }
    //Draw component first, then draw this decoration
    public void draw(Graphics2D g2d)
    {
        if (decorator != null) decorator.draw(g2d);
        drawThis(g2d);
    }
    abstract public void drawThis(Graphics2D g2d);
}
```

DrawAxisTicks

The DrawAxisTicks class reads

```
package graphicsDomain;

import java.awt.*;
import java.awt.geom.*;
import java.awt.font.*;
import java.text.*;
import java.util.*;
// Encapsulates methods to draw axes and tick marks and
// labels
public abstract class DrawAxisTicks extends GraphDecorator
{
    private Font tickFont;
    protected double textFac;
                    //(ascent+leading)/(text-height)
    private double maxNormalRange = 10;
    private double minNormalRange = 1;
                // (Do not use sci notation inside
                //   of normal range)
    private int axisLen; //Length in pixels of the axis

//protected variables, will be used by subclasses
    protected double
        currentTickVal, //Current tick value (either X and Y)
        stringWidth, //Witdh of the current tick label
        stringHeight; //Height of the current tick label
    protected int
        borderSize, //Border size in pixels
        xSize, //Graph panel width
        ySize, //Graph panel height
        tickPix, //Tick size in pixels
        axisLabelHeight, //height of the current label
        currentTickPixel;
                //Pixel position of the current tick
    protected String tickLabel;
                //The label associated with the current tick

    //Nice formater for axis numbers
    private NumberFormat tickFormatter =
                    NumberFormat.getNumberInstance();

    public DrawAxisTicks(GraphMediator med) { super(med);}
//Draw generic axis ticks. The method will call specific
//methods implemented by subclasses for each axis
    public void drawThis(Graphics2D g2D)
    {
        g2D.setColor(med.getGraphOptions().
            selectColor(med.getGraphOptions().getLineColor()));
        g2D.setStroke(med.getGraphOptions().getAxisStroke());
```

```java
//Get information from GraphOptionsInGraphics
//instance
tickFont = med.getGraphOptions().getTickFont();
textFac = med.getGraphOptions().getTextFac();
maxNormalRange = med.getGraphOptions().
                             getMaxNormalRange();
minNormalRange = med.getGraphOptions().
                             getMinNormalRange();
int axisLabelHeight = med.getGraphOptions().
                             getAxisFontHeight();
int sigFigs = med.getGraphOptions().getSigFigs();
tickPix = med.getGraphOptions().getTickPix();
xSize = med.getXSize();
ySize = med.getYSize();
borderSize = med.getBorderSize();
double[] maxMinLen = getMaxMinLen();
double dataMin = maxMinLen[0];
double dataMax = maxMinLen[1];
axisLen = (int) maxMinLen[2];

// set font and find FontRenderContext for
//label dimensions
g2D.setFont(tickFont);
FontRenderContext context =
                     g2D.getFontRenderContext();

String dummyTickL = ".";
              //compute width of dummy tick label
for (int i = 0; i < sigFigs; i++)
{
    dummyTickL = dummyTickL + "W";
}
Rectangle2D bounds = tickFont.getStringBounds(
                             dummyTickL, context);
int textWidth = (int) bounds.getWidth();

//find number of ticks
int numTicks = (int) (
               (double)(axisLen - 2 * borderSize) /
               (double) textWidth);
if (numTicks < 4) numTicks = 4;
int nBin = numTicks - 1;

double dataRange, d;
if (med.getGraphOptions().getNiceNumbers())
{
    dataRange = nicenum(dataMax - dataMin, true);
    d = nicenum(dataRange / (nBin), false);
```

```
    }
    else
    {
        dataRange = dataMax - dataMin;
        d = dataRange / nBin;
    }

    // tick range should be smaller than the data range
    double tickMin = Math.ceil(dataMin / d) * d;
    double tickMax = Math.floor(dataMax / d) * d;

    // Fill in an arraylist of tick values
    ArrayList tickPoints = new ArrayList();
    for (double t = tickMin; t<=tickMax + 0.5 * d;
        t = t + d)
    {
        tickPoints.add(new Double(t));
    }

    Double dummyMinTick = (Double) tickPoints.get(0);
                                                   //first
    double minTick = dummyMinTick.doubleValue();
    Double dummyMaxTick = (Double) tickPoints.get(
                    tickPoints.size() - 1); //last
    double maxTick = dummyMaxTick.doubleValue();

    // Determine whether to use scientific notation;
    // find multiplier and sciLabel
    double maxAbsTick = Math.max(Math.abs(maxTick),
                                Math.abs(minTick));
    int powerOfTen = 0;
    double multiplier = 1;
    boolean scientific = false;
    String sciLabel = "";
    if (maxAbsTick >= maxNormalRange ||
                        maxAbsTick < minNormalRange)
    {
        if(maxAbsTick == 0)
            powerOfTen = 1;
        else
            powerOfTen = (int) Math.floor(
                            Math.log10(maxAbsTick));
        scientific = true;
        sciLabel = "   (x10^" + powerOfTen + ")";
        multiplier = Math.pow(10, powerOfTen);
    }
    for (int i = 0; i < tickPoints.size(); i++)
    {
        Double currVal = (Double) tickPoints.get(i);
```

```java
            currentTickVal = currVal.doubleValue();

            // draw little tick mark and grid line.
            //This is done by subclass method
            currentTickPixel = drawTickGrid(g2D);

            // Find tick label
            tickLabel = "";
            double tickNumber = 0.;
            if (scientific)
            {
                tickNumber = currentTickVal / multiplier;
            } // Divide by exponent to get number
            else
            {
                tickNumber = currentTickVal;
            }
            tickLabel = tickFormatter.format(tickNumber);

            // Calculate tick label width and height
            bounds = tickFont.getStringBounds(
                                    tickLabel, context);
            stringWidth = bounds.getWidth();
            stringHeight = bounds.getHeight();
            // draw tick label. Will be carried out by
            // subclass method.
            drawTickLabel(g2D);
        }
        //The label specifying the tick values multipliers
        //needs to be appended to
        //either X or Y labels stored
        //in the GraphDataInGraphics instance.
        //This is done in the concrete subclasses.
        setSciLabel(sciLabel);
    }
    //Get maximum, minumum avlues and size in pixel
    //(returned in a double array)
    //for either X or Y axis
    public abstract double[] getMaxMinLen();
    //Draw the current line in the tick grid (either X or Y),
    //based on the current tick value
    public abstract int drawTickGrid(Graphics2D g2D);
    //Draw the current tick label (either X or Y)
    public abstract void drawTickLabel(Graphics2D g2D);
    //Store x or y scientific label suffix in med
    public abstract void setSciLabel(String label);
    protected double nicenum(double x, boolean round)
    {
        ...
```

```
        // as for earlier lisiting of nicenum
        ...
}
```

DrawAxisTicksX

The `DrawAxisTicksX` class reads:

```
package graphicsDomain;

import java.awt.*;
/** Encapsulates methods to draw axes and tick marks and
    labels */
public class DrawAxisTicksX extends DrawAxisTicks
{
    public DrawAxisTicksX(GraphMediator med) { super(med);}
    //Get maximum, minumum avlues and size in pixel
    //(returned in a double array)
    //for either X axis
    public double[] getMaxMinLen()
    {
        double xMin = med.getMinXData();
        double xMax = med.getMaxXData();
        double dLen = med.getXSize();
        return new double[] {xMin,xMax,dLen};
    }
    //Draw the current line in the X tick grid   based
    //on the current tick value
    public int drawTickGrid(Graphics2D g2D)
    {
        // Draw little tick mark
        int gp = med.getXPixel(currentTickVal);
        int yu = (int) ySize-borderSize;
        int yl = yu+tickPix;
                // Number of pixels below the axis for tick
        g2D.drawLine(gp,yu,gp,yl); // Draw little tick

        // Draw grid line
        g2D.drawLine(gp,yu,gp,borderSize);
        return gp;
    }
    //Draw the current X tick label
    public void drawTickLabel(Graphics2D g2D)
    {
        // Draw label
        int xp = currentTickPixel - (int) (stringWidth/2.0);
        int tickOffset = 3;
        int yp = (int) (ySize - borderSize + tickOffset
```

```java
                                    + textFac*stringHeight );
        if ( borderSize<stringHeight )
            yp = (int) ( ySize - stringHeight +
                                    textFac*stringHeight );
        g2D.drawString( tickLabel , xp , yp );
    }
    //Store X scientific label suffix in med
    public void setSciLabel( String label )
    {
        med.getGraphData().setXSuffix( label );
    }
    public String toString()
    {
        return "DrawAxisTicksX";
    }
}
```

DrawAxisTicksY

The `DrawAxisTicksY` class is similar to the `DrawAxisTicksX` class with the exception that the ticks will be drawn horzontally and the label will be rotated. The `drawTickLabel` method is listed below:

```java
package graphicsDomain;

import java.awt.*;
/** Encapsulates methods to draw axes and tick marks and
    labels */
public class DrawAxisTicksY extends DrawAxisTicks
{
    public DrawAxisTicksY( GraphMediator med ) { super(med); }
    //Get maximum, minumum avlues and size in pixel
    //( returned in a double array )
    //for Y axis
    public double[] getMaxMinLen()
    {
        double yMin = med.getMinYData();
        double yMax = med.getMaxYData();
        double dLen = med.getYSize();
        return new double[] { yMin , yMax , dLen };
    }
    //Draw the current line in the Y tick grid ,
    //based on the current tick value
    public int drawTickGrid( Graphics2D g2D )
    {
        // Draw little tick mark
        int gp = med.getYPixel( currentTickVal );
```

```java
        int xr = (int) borderSize;
        int xl = xr-tickPix;
                // Number of pixels left of the axis for tick
        g2D.drawLine(xl,gp,xr,gp); // Draw little tick

        // Draw grid line
        g2D.drawLine(borderSize,gp,(int)xSize-borderSize,gp);
        return gp;
    }
    //Draw the current Y tick label
    public void drawTickLabel(Graphics2D g2D)
    {
        // Draw label
        int xp = (int) (1.5*axisLabelHeight +
                                    textFac*stringHeight);
        int yp = currentTickPixel + ((int)(stringWidth/2.0));
        // Rotate and draw string; rotate back
        g2D.rotate(-90.0*Math.PI/180.,xp,yp);
        g2D.drawString(tickLabel,xp,yp);
        g2D.rotate(90.0*Math.PI/180.,xp,yp);
    }
    //Store Y scientific label suffix in med
    public void setSciLabel(String label)
    {
        med.getGraphData().setYSuffix(label);
    }
    public String toString()
    {
        return "DrawAxisTicksY";
    }
}
```

DrawCentredMessage

The `DrawCentredMessage` class reads:

```java
package graphicsDomain;

import java.awt.*;
import java.awt.geom.*;
import java.awt.font.*;

//Encaplulates methods for drawing a centered message.
//Used to display error
//messages in Graph panels
public class DrawCentredMessage extends GraphDecorator
{
    String message;
```

```java
    public DrawCentredMessage(GraphMediator med)
                                        { super(med); }

    public void drawThis(Graphics2D g2)
    {
        //Set appearance as defined in
        //GraphOptionsInGraphics instance
        g2.setColor(med.getGraphOptions().selectColor(
            med.getGraphOptions().getTitleColor()));
        g2.setStroke(med.getGraphOptions().getTitleStroke());

        g2.setFont(med.getGraphOptions().getTitleFont());

        FontRenderContext context =
            g2.getFontRenderContext();
        Rectangle2D bounds =
            med.getGraphOptions().getTitleFont().
            getStringBounds(message, context);
        double stringWidth = bounds.getWidth();
        double stringHeight = bounds.getHeight();

        //Draw Centered message
        int xp = (int) ((med.getXSize() -
                        stringWidth) / 2.0);
        int yp = (int) (med.getYSize() / 2. +
                (med.getBorderSize() - stringHeight) / 2.0);
        g2.drawString(message, xp, yp);
    }

    public void setMessage(String message)
    {
        this.message = message;
    }
    public String toString()
    {
        return "DrawCentredMessage";
    }
}
```

DrawCrossHair

The `DrawCrossHair` class reads:

```java
package graphicsDomain;

import java.awt.*;
```

```java
import java.awt.geom.*;

//Encapsulates methods for drawing a crosshair centered
//on the waveform at the given X position
public class DrawCrossHair    extends GraphDecorator
{
    //The current mouse X position is used to find the
    //centre of the crosshair
    private int currX;

    public DrawCrossHair(GraphMediator med) { super(med);}

    public void setCurrX(int currX)
    {
        this.currX = currX;
    }

    //Draw the crosshair
    public void drawThis(Graphics2D g2)
    {
        Point[] points = med.getPixelArray(
            med.getGraphData().getX(),
            med.getGraphData().getY());
        int xSize = med.getXSize();
        int ySize = med.getYSize();

        /* Draw cursor */
        // First find index of closest x data point
        // find closest data value from xScaled
        int minIdx = 0;
        double minDist = Math.abs(currX - points[0].x);
        double currDist = minDist;
        for (int i = 1; i < points.length; i++)
        {
            currDist = Math.abs(currX - points[i].x);
            if (currDist < minDist)
            {
                minDist = currDist;
                minIdx = i;
            }
        }
        // set xData to xVals at minimum distance
        double xdata = med.getXValue(points[minIdx].x);
        double ydata = med.getYValue(points[minIdx].y);

        //Notify listener of the update
        med.getGraphicsFacade().graphPointUpdated(
                                        xdata, ydata, true);
        // draw large cursor marker on graph
```

```
            g2.setStroke(med.getGraphOptions().getLineStroke());
            Line2D xLine = new Line2D.Float(0, points[minIdx].y,
                                            xSize, points[minIdx].y);
            Line2D yLine = new Line2D.Float(points[minIdx].x,
                                            ySize, points[minIdx].x, 0);
            g2.draw(xLine);
            g2.draw(yLine);
    }
    public String toString()
    {
            return "DrawCrossHair";
    }
}
```

DrawLabels

The `DrawLabels` class reads:

```
package graphicsDomain;

import java.io.*;
import java.awt.*;
import java.awt.event.*;
import javax.swing.*;
import javax.swing.event.*;
import javax.swing.tree.*;
import java.awt.geom.*;
import java.awt.font.*;
import java.text.*;
import java.util.*;

public class DrawLabels  extends GraphDecorator
{
    public DrawLabels(GraphMediator med) { super(med);}

    public void drawThis(Graphics2D g2)
    {
        g2.setColor(med.getGraphOptions().selectColor(
            med.getGraphOptions().getTitleColor()));
        g2.setStroke(med.getGraphOptions().getTitleStroke());

        int xSize = med.getXSize();
        int ySize = med.getYSize();
        int borderSize = med.getBorderSize();
        double textFac = med.getGraphOptions().getTextFac();

        // Graph Title
```

```java
        Font titleFont=med.getGraphOptions().getTitleFont();
        g2.setFont(titleFont);//find size of label
        String title = med.getGraphData().getTitle();
        FontRenderContext context =
                            g2.getFontRenderContext();
        Rectangle2D bounds = titleFont.getStringBounds(title,
                                                 context);
        double stringWidth = bounds.getWidth();
        double stringHeight = bounds.getHeight();

        int xp = (int)((xSize - stringWidth)/2.0);
        int yp = (int)((borderSize - stringHeight)/2.0 +
                                    textFac*stringHeight);
        if (borderSize<stringHeight)
            yp = (int)(textFac*stringHeight);
        g2.drawString(title, xp, yp);

        Font axisFont = med.getGraphOptions().getAxisFont();
        g2.setFont(axisFont);
        /* Draw X-axis label */
        context = g2.getFontRenderContext();
        String xLabel = med.getGraphData().getFullXLabel();
        bounds =
            axisFont.getStringBounds(xLabel, context);
        stringWidth = bounds.getWidth();
        stringHeight = bounds.getHeight();

        xp = (int)((xSize - stringWidth)/2.0);
        yp = (int)(ySize - 1.25*stringHeight +
                                  textFac*stringHeight);
        g2.drawString(xLabel, xp, yp);

        /* Draw Y-axis label */
        String yLabel = med.getGraphData().getFullYLabel();
        bounds = axisFont.getStringBounds(yLabel, context);
        stringWidth = bounds.getWidth();
        stringHeight = bounds.getHeight();

        xp = (int)(stringHeight*0.5+ stringHeight*textFac);
        yp = (int)(ySize - (ySize - stringWidth)/2.0);
        // rotate graphics context; draw the message;
        // rotate back
        g2.rotate(-90.0*Math.PI/180.,xp,yp);
        g2.drawString(yLabel, xp, yp);
        g2.rotate(+90.0*Math.PI/180.,xp,yp);
    }

    public String toString()
    {
```

```
            return "DrawLabels";
    }
}
```

DrawWaveform

The `DrawWaveform` class reads:

```
package graphicsDomain;

import java.awt.*;
import java.awt.geom.Line2D;

public class DrawWaveform extends GraphDecorator
{
    public DrawWaveform(GraphMediator med) { super(med); }

    public void drawThis(Graphics2D g2)
    {
        Point[] points = med.getPixelArray(
            med.getGraphData().getX(),
            med.getGraphData().getY());
        g2.setStroke(med.getGraphOptions().getLineStroke());
        g2.setColor(med.getGraphOptions().selectColor(
            med.getGraphOptions().getLineColor()));

        int xSize = med.getXSize();
        for (int i = 0; i < points.length - 1; i++)
        {
            if (points[i + 1].x < 0 || points[i].x > xSize)
                continue;
            Line2D line = new Line2D.Float(
                        points[i].x, points[i].y,
                        points[i + 1].x, points[i + 1].y);
            g2.draw(line);

        }
    }
    public String toString()
    {
        return "DrawWaveform";
    }
}
```

DrawZoomBox

The `DrawZoomBox` class reads:

```java
package graphicsDomain;

import java.awt.*;

//Encapsulates methods for drawing a zoom box
public class DrawZoomBox  extends GraphDecorator
{
    //Pixel coordinates of the zoom box rectangle
    private int startX, endX, startY, endY;

    //Zooming flag
    private boolean isZooming = false;

    public DrawZoomBox(GraphMediator med) { super(med);}

    //Accessor methods
    public void setStartX(int startX)
    {
        this.startX = startX;
    }
    public void setEndX(int endX)
    {
        this.endX = endX;
    }
    public void setStartY(int startY)
    {
        this.startY = startY;
    }
    public void setEndY(int endY)
    {
        this.endY = endY;
    }
    public void setZooming(boolean isZooming)
    {
        this.isZooming = isZooming;
    }

    //Draw implementation
    public void drawThis(Graphics2D g2)
    {
        //If not zooming, no zoom box is drawn
        if(!isZooming) return;
        //Otherwise draw the zoom box
        g2.setStroke(med.getGraphOptions().getLineStroke());
        g2.drawRect(startX, startY, endX - startX,
                endY - startY);
    }
    public String toString()
```

```
            {
                return "DrawZoomBox";
            }

}
```

E.5.2 Adapter Classes

`GraphDataInGraphics`

The `GraphDataInGraphics` class reads:

```
package graphicsDomain;
import sharedInterfaces.*;
import sharedDataInterfaces.*;

/** Adapted version of GraphData.
Adds scientific suffix to axes labels. Can set special
values of minYData, maxYData, etc., to guard against
divide by zero.*/
public class GraphDataInGraphics
{
    private GraphDataInterface graphData;
    private String xSuffix, ySuffix;
    double minXData=0,maxXData=0,minYData=0,maxYData=0;

    public GraphDataInGraphics(GraphDataInterface gData)
    {
        graphData=gData;
        if (graphData.getIsGraphData())
        {
            double xMin = graphData.getMinX();
            double xMax = graphData.getMaxX();
            double yMin = graphData.getMinY();
            double yMax = graphData.getMaxY();
            double deltaX = xMax - xMin;
            double deltaY = yMax - yMin;

            // Deal with small denominators
            if (Math.abs(deltaY)<1.e-6)
                                // warning: hard coding
            { // Cope with zero range in the data
                System.out.println(
                        " GraphDataInGraphics: "+
                        "small demoninator in y values");
                double absVal = Math.abs(yMin);
                if (absVal > 0.)
                {
```

```java
                    yMin = yMin - 0.5*absVal;
                    yMax = yMax + 0.5*absVal;
                }
                else
                {
                    yMin = -.5;
                    yMax = .5;
                }
            }
            if (Math.abs(deltaX)<1.e-6)
                                // warning: hard coding
            { // Cope with zero range in the data
                System.out.println(
                        " GraphDataInGraphics: "+
                        "small demoninator in x values");
                double absVal = Math.abs(xMin);
                if (absVal > 0.)
                {
                    xMin = xMin - 0.5*absVal;
                    xMax = xMax + 0.5*absVal;
                }
                else
                {
                    xMin = -.5;
                    xMax = .5;
                }
            }
            minXData = xMin;
            maxXData = xMax;
            minYData = yMin;
            maxYData = yMax;
        }
    }
    public void setXSuffix(String suffix)
    {
        xSuffix = suffix;
    }
    public void setYSuffix(String suffix)
    {
        ySuffix = suffix;
    }
    public String getFullXLabel()
    {
        return graphData.getXLabel() + xSuffix;
    }
    public String getFullYLabel()
    {
        return graphData.getYLabel() + ySuffix;
    }
```

```
        public int getLength(){ return graphData.getLength();}
        public double getMinX(){ return minXData;}
        public double getMaxX(){ return maxXData;}
        public double getMinY(){ return minYData;}
        public double getMaxY(){ return maxYData;}
        public double[] getX(){ return graphData.getX();}
        public double[] getY(){ return graphData.getY();}
        public String getTitle(){ return graphData.getTitle();}
        public boolean getIsGraphData()
                        { return graphData.getIsGraphData();}
        public boolean getIsDataError()
                        { return graphData.getIsDataError();}
}
```

GraphOptionsInGraphics

The GraphOptionsInGraphics class reads:

```
package graphicsDomain;
import sharedDataInterfaces.GraphOptionsInterface;

import java.awt.*;
import java.awt.font.*;

/** Adapted class. Adds various graph parameters to
    those returned from
    the GraphOptionsDialog.*/
public class GraphOptionsInGraphics
{
    private GraphOptionsInterface graphOptions;
    private int sigFigs, titleFontHeight,
            axisFontHeight, tickFontHeight, tickPix;
    private double textFac;
    private Font titleFont, axisFont, tickFont;
    private double maxNormalRange=10;
    private double minNormalRange=1;
                    // (Do not use sci notation inside
                    //  of normal range for ticks)

    private BasicStroke lineStroke, axisStroke, titleStroke;

    public GraphOptionsInGraphics(
                        GraphOptionsInterface options)
    {
        graphOptions = options;
        sigFigs = 3;
        titleFontHeight = 12;
        axisFontHeight = 10;
```

```
        tickFontHeight = 10;
        textFac = 0.8; // (ascent+leading)/(text−height)
        tickPix = 4; // number of pixels for tick length
        titleFont = new Font(null, Font.PLAIN,
                                    titleFontHeight);
        axisFont = new Font(null, Font.PLAIN,
                                    axisFontHeight);
        tickFont = new Font(null, Font.PLAIN,
                                    tickFontHeight);
        // stroke for waveform
        float strokeWidth = 1.0f;
        float[] solid = {12.0f,0.0f};    // Solid line style
        float[] dashed = {1.0f,5.0f};    // Dashed line style
        lineStroke = new BasicStroke( strokeWidth,
                    BasicStroke.CAP_SQUARE,
                    BasicStroke.JOIN_MITER, 1.0f,
                    solid, 0.0f );
        // stroke for axes and ticks
        strokeWidth = 0.25f;
        axisStroke = new BasicStroke( strokeWidth,
                    BasicStroke.CAP_SQUARE,
                    BasicStroke.JOIN_MITER, 1.0f,
                    dashed, 0.0f );
        titleStroke = axisStroke;
    }
    public void updateOptions(GraphOptionsInterface opt)
    {   graphOptions=opt;}
    public void setTickPix(int tp)
    {   tickPix = tp;}
    public int getSigFigs(){ return sigFigs;}
    public int getTitleFontHeight(){ return titleFontHeight;}
    public int getAxisFontHeight(){ return axisFontHeight;}
    public int getTickFontHeight(){ return tickFontHeight;}
    public int getTickPix(){ return tickPix;}
    public double getTextFac(){ return textFac;}
    public Font getTitleFont(){ return titleFont;}
    public Font getAxisFont(){ return axisFont;}
    public Font getTickFont(){ return tickFont;}
    public BasicStroke getLineStroke() { return lineStroke;}
    public BasicStroke getAxisStroke() { return axisStroke;}
    public BasicStroke getTitleStroke()
                            { return titleStroke;}
    public double getMaxNormalRange()
    {
        return maxNormalRange;
    }
    public double getMinNormalRange()
    {
        return minNormalRange;
```

```java
    }
    // methods delegated to GraphOptions
    public int getTitleColor()
    {    return graphOptions.getTitleColor();}
    public int getGraphColor()
    {    return graphOptions.getGraphColor();}
    public int getLineColor()
    {
        return graphOptions.getLineColor();
    }
    public Color selectColor(int i)
    {    return graphOptions.selectColor(i);}
    public Color getTickLineColor()
    {
        return Color.GRAY;
    }
    public String selectColorName(int i)
    {    return graphOptions.selectColorName(i);}
    public boolean getDefaultBorder()
                {return graphOptions.getDefaultBorder();}
    public boolean getNiceNumbers()
                {return graphOptions.getNiceNumbers();}
    public int getDisplayMode()
                {return graphOptions.getDisplayMode();}
    public int getZOOM()
                { return graphOptions.ZOOM_MODE;}
    public int getCROSSHAIR()
                { return graphOptions.CROSSHAIR_MODE;}
    public int getGRAB() { return graphOptions.GRAB_MODE;}
}
```

E.5.3 GraphMediator

The `GraphMediator` class reads

```java
package graphicsDomain;

import java.awt.*;
import java.awt.font.*;
import java.awt.geom.*;
import java.awt.event.*;
import javax.swing.*;
import java.util.*;
import java.lang.*;
import java.text.*;
/** Performs conversions between world and pixel
    coordinates */
public class GraphMediator
```

```
{
    private GraphicsFacade graphFacade;
    private GraphDataInGraphics graphData;
    private GraphOptionsInGraphics graphOptions;
    private double minXData, maxXData, minYData, maxYData;
    private int borderSize, xSize, ySize, currX=0, currY=0;
    private String message="";
                // possible text message to draw
    private Point[] scaledPoints;
    private boolean isGraphData, isDataError;

    private GraphDecorator decorator=null;
    private DrawAxisTicksX axisXDecorator;
    private DrawAxisTicksY axisYDecorator;
    private DrawWaveform waveformDecorator;
    private DrawCrossHair crossHairDecorator;
    private DrawCentredMessage warningDecorator;
    private DrawLabels labelsDecorator;
    private DrawZoomBox zoomDecorator;

    public GraphMediator(GraphicsFacade g)
    {
        graphFacade = g;
        graphData = null;
        graphOptions = null;
        isGraphData = false;
        isDataError = false;
        //Instantiate decorators
        warningDecorator = new DrawCentredMessage(this);
        zoomDecorator = new DrawZoomBox(this);
        waveformDecorator = new DrawWaveform(this);
        crossHairDecorator = new DrawCrossHair(this);
        labelsDecorator = new DrawLabels(this);
        axisXDecorator = new DrawAxisTicksX(this);
        axisYDecorator = new DrawAxisTicksY(this);
    }
    public void updateData(GraphDataInGraphics gD)
    {
        graphData=gD;
        update();
        createDecorator();
    }
    public void updateOptions(GraphOptionsInGraphics gO)
    {
        graphOptions = gO;
        createDecorator();
    }
    public void update()
    {
```

```java
        if (graphData!=null) //is null for empty constructor
        {
            minXData=graphData.getMinX();
            maxXData=graphData.getMaxX();
            minYData=graphData.getMinY();
            maxYData=graphData.getMaxY();
            isGraphData = graphData.getIsGraphData();
            isDataError = graphData.getIsDataError();
            if (maxXData <= minXData)
                minXData = maxXData - 1E-20;
                            //Guard against divide by zero
        }
    }
    public void createDecorator()
    {
        /*Build the decorator chain. The elements in the
          chain define
          what is going to be displayed.

          If no data or bad data then the decorator
          chain is only composed
          of a DrawCentredMessage object
          containing an error message.*/
        if (graphData == null || !isGraphData)
        {
            if (graphData != null && isDataError)
                warningDecorator.setMessage(
                    "Unexpected data. See console message.");
            else
                warningDecorator.setMessage("No data");

            //Build decorator chain with a centred message
            decorator = warningDecorator;
        }
        else //Build the decorator chain based on
             //the current display mode
        {
            if (graphOptions.getDisplayMode()==
                            graphOptions.getCROSSHAIR())
                //decorator chain with X and Y axis,
                //crosshair and waveform
                decorator = concatenateDecorators(
                        waveformDecorator,
                        labelsDecorator,
                        crossHairDecorator,
                        axisYDecorator,
                        axisXDecorator);

            if (graphOptions.getDisplayMode()==
```

```java
                        graphOptions.getZOOM())
            //decorator chain with X and Y axis,
            //zoom box and waveform
            decorator = concatenateDecorators(
                    waveformDecorator,
                    labelsDecorator,
                    zoomDecorator,
                    axisYDecorator,
                    axisXDecorator);

        if (graphOptions.getDisplayMode()==
                        graphOptions.getGRAB())
            //decorator chain with X and Y axis and
            //waveform
            decorator = concatenateDecorators(
                    waveformDecorator,
                    labelsDecorator,
                    axisYDecorator,
                    axisXDecorator);
    }
}
/** Forms decorator chain. Note variable argument list.*/
private GraphDecorator concatenateDecorators(
                        GraphDecorator ... decorators)
{
    for (int i = decorators.length − 2; i >= 0; i−−)
        decorators[i].setDecorator(decorators[i + 1]);
    return decorators[0];
}
public GraphDecorator getDecorator()
{
    return decorator;
}
// Mappings from world to pixel coordinates.
/** World to pixel in X*/
public int getXPixel(double xVal)
{
    return (int) ( (xVal−minXData)/(maxXData−minXData)
            * (xSize −2*borderSize) + borderSize );
}
/** World to pixel in Y*/
public int getYPixel(double yVal)
{
    return (int) ( (yVal−minYData)/(maxYData−minYData)
            * (2*borderSize−ySize) + ySize−borderSize );
}
/** Pixel to world in X*/
public double getXValue(double xPixel)
{
```

```java
            return ( (xPixel-borderSize)/(xSize -2*borderSize)
                    * (maxXData-minXData) + minXData );
}
/** Pixel to world in Y*/
public double getYValue(double yPixel)
{
    return ( (yPixel-ySize + borderSize)/
            (2*borderSize - ySize)
            * (maxYData-minYData) + minYData );
}
/** Returns an array of (possibly filtered)
    pixel Point objects */
public Point[] getPixelArray(double []xVal, double[]yVal)
{
  int actDim = -1;
  int currXVal, currYVal, prevXVal = 0;
  for (int i = 0; i < yVal.length; i++)
  {
                //Find how many points can be filtered
    currXVal = (int) ( (xVal[i] - minXData) /
            (maxXData - minXData)
            * (xSize - 2 * borderSize) + borderSize);
    if (actDim == -1 || currXVal != prevXVal) {
      actDim++;
            // increment counter if x pixel value is
            // different
      prevXVal = currXVal;
    }
  }
  if (actDim > yVal.length / 2)
  { //Not enough points to bother filtering
    scaledPoints = new Point[yVal.length];
    for (int i = 0; i < yVal.length; i++) {
      scaledPoints[i] = new Point();
      scaledPoints[i].x = (int) ( (xVal[i] - minXData) /
            (maxXData - minXData)
            * (xSize - 2 * borderSize) + borderSize);
      scaledPoints[i].y = (int) ( (yVal[i] - minYData) /
            (maxYData - minYData)
            * (2 * borderSize - ySize)
            + ySize - borderSize);
    }
    return scaledPoints;
  }
  else { //It makes sense to filter.
        //Adjacent filtered points store max/min Y vals
    scaledPoints = new Point[2 * (actDim + 1)];
    actDim = -1;
    for (int i = 0; i < yVal.length; i++) {
```

```java
            currXVal = (int)
                    ( (xVal[i] - minXData) / (maxXData - minXData)
                     * (xSize - 2 * borderSize) + borderSize);
            currYVal = (int)
                    ( (yVal[i] - minYData) / (maxYData - minYData)
                     * (2 * borderSize - ySize)
                     + ySize - borderSize);
            if (actDim == -1 || currXVal != prevXVal) {
              actDim++;
              prevXVal = currXVal;
              scaledPoints[2 * actDim] =
                            new Point(currXVal, currYVal);
              scaledPoints[2 * actDim + 1] =
                            new Point(currXVal, currYVal);
            }
            else { // store max/min y vals
              if (currYVal < scaledPoints[2 * actDim].y)
                 scaledPoints[2 * actDim].y = currYVal;
              if (currYVal > scaledPoints[2 * actDim + 1].y)
                 scaledPoints[2 * actDim + 1].y = currYVal;

            }
        }
        return scaledPoints;
    }
}
public void setXSize(int x) {xSize=x;}
public void setYSize(int y) {ySize=y;}
public void setBorderSize(int b){borderSize=b;}
public double getMinXData() { return minXData;}
public double getMaxXData() { return maxXData;}
public double getMinYData() { return minYData;}
public double getMaxYData() { return maxYData;}
public void setMinXData(double minXData)
                              { this.minXData = minXData;}
public void setMaxXData(double maxXData)
                              { this.maxXData = maxXData;}
public void setMinYData(double minYData)
                              { this.minYData = minYData;}
public void setMaxYData(double maxYData)
                              { this.maxYData = maxYData;}
public int getXSize() {return xSize;}
public int getYSize() {return ySize;}
public int getBorderSize(){return borderSize;}
public boolean getIsGraphData() {return isGraphData;}
public boolean getIsDataError() {return isDataError;}
public GraphicsFacade getGraphicsFacade()
                              { return graphFacade;}
public GraphOptionsInGraphics getGraphOptions()
```

```
                              {return graphOptions;}
    public GraphDataInGraphics getGraphData()
                              {return graphData;}
    public DrawCrossHair getCrossHairDecorator()
                              { return crossHairDecorator;}
    public DrawZoomBox getZoomDecorator()
                              { return zoomDecorator;}
    /** Message string for writing on graph */
    public String getMessage() { return message;}
    public void setMessage(String _message)
                              { message = _message;}
    /** Current XY positions of cursor */
    public void setCurrX(int xPixel) { currX = xPixel;}
    public int getCurrX() { return currX;}
    public void setCurrY(int yPixel) { currY = yPixel;}
    public int getCurrY() { return currY;}
}
```

E.5.4 GraphicsFacade

The `GraphicsFacade` class reads:

```
package graphicsDomain;

import sharedInterfaces.*;
import sharedDataInterfaces.*;
import java.awt.*;
import java.awt.event.*;
import java.awt.geom.*;
import java.awt.font.*;
import java.util.*;

public class GraphicsFacade extends AbstractGraphicsFacade
{
    private GraphMediator med;
    private GraphMouseHandler mouseHandler =
                              new GraphMouseHandler();
    private GuiFacadeInterface guiFacade;
                //Used to draw current waveform coords
    public GraphicsFacade()
    {
        med = new GraphMediator(this);
        addMouseListener(mouseHandler);
        addMouseMotionListener(mouseHandler);
    }
    public void initialise(GuiFacadeInterface guiFacade)
    {
        this.guiFacade = guiFacade;
```

```java
}
public void setGraphData(GraphDataInterface data)
{
    GraphDataInGraphics graphData =
                    new GraphDataInGraphics(data);
    med.updateData(graphData);
}
public void setGraphOptions(
                    GraphOptionsInterface graphOpt)
{
    GraphOptionsInGraphics options =
                    new GraphOptionsInGraphics(graphOpt);
    med.updateOptions(options);
}
public void applyGraphOptions()
{
    med.createDecorator();
    repaint();
}
public void graphPointUpdated(double x,
                    double y, boolean inRange)
{
    guiFacade.graphPointUpdated(x, y, inRange);
}
public void paintComponent(Graphics g)
{
    // Cast the graphics object to Graph2D
    Graphics2D g2 = (Graphics2D) g;
    // Get plot size
    Dimension size = getSize();
    int xSize = size.width;
    int ySize = size.height;
    // find border size
    int borderSize;
    if (med.getGraphOptions().getDefaultBorder())
    {
        borderSize = calcBorderSize(g2);
        med.getGraphOptions().setTickPix(4);
    }
    else
    {
        borderSize = 0;
        med.getGraphOptions().setTickPix(0);
    }
    // Set graph and border size in med
    med.setXSize(xSize);
    med.setYSize(ySize);
    med.setBorderSize(borderSize);
    // Set background Color
```

```java
        g2.setColor(med.getGraphOptions().selectColor(
            med.getGraphOptions().getGraphColor()));
        g2.fill(new Rectangle2D.Double(0, 0, xSize, ySize));
        //call draw method of the decorator chain
        med.getDecorator().draw(g2);
    }
    public int calcBorderSize(Graphics2D g2)
    {
        /* borderSize calculation using test string "Test"*/
        double border1, border2, border3, spaceFac = 1.5;
        // height of top adornment: graph title
        Font titleFont =
                med.getGraphOptions().getTitleFont();
        g2.setFont(titleFont); // title font
        FontRenderContext context =
                g2.getFontRenderContext();
        Rectangle2D bounds = titleFont.getStringBounds(
                                        "Test", context);
        double stringHeight = bounds.getHeight();
        border1 = spaceFac * stringHeight;
        // height of bottom adornments: x-axis label
        //plus x-tick-labels
        Font axisFont = med.getGraphOptions().getAxisFont();
        g2.setFont(med.getGraphOptions().getAxisFont());
                                        // axis font
        context = g2.getFontRenderContext();
        bounds = axisFont.getStringBounds("Test", context);
        stringHeight = bounds.getHeight();
        border2 = spaceFac * stringHeight;
        Font tickFont = med.getGraphOptions().getTickFont();
        g2.setFont(tickFont); // tick font
        context = g2.getFontRenderContext();
        bounds = tickFont.getStringBounds("Test", context);
        stringHeight = bounds.getHeight();
        border3 = spaceFac * stringHeight;
        return (int) Math.max(border1, border2 + border3);
    }
    /** inner class to handle mouse events */
    private class GraphMouseHandler
        implements MouseListener, MouseMotionListener
    {
        //Recorded mouse position when mouse button pressed
        private int startX = 0, startY = 0;
        //Recorded previous mouse position
        private int prevX = 0, prevY = 0;
        public void mousePressed(MouseEvent e)
        {
            //Record mouse position when clocked
            startX = prevX = e.getX();
```

```java
            startY = prevY = e.getY();
            switch (med.getGraphOptions().getDisplayMode())
            {
                //Notify decorators interested in current
                //mouse position: zoomDecorator and
                //crossHairDecorator
                case GraphOptionsInterface.GRAB_MODE:
                {
                    GraphicsFacade.this.repaint();
                    break;
                }
                case GraphOptionsInterface.CROSSHAIR_MODE:
                {
                    med.getCrossHairDecorator().
                                    setCurrX(startX);
                    GraphicsFacade.this.repaint();
                    break;
                }
                case GraphOptionsInterface.ZOOM_MODE:
                {
                    med.getZoomDecorator().setStartX(startX);
                    med.getZoomDecorator().setStartY(startY);
                    med.getZoomDecorator().setZooming(true);
                    break;
                }
            }
        }
        public void mouseMoved(MouseEvent e)
        {   //define cursor shapes for the different modes
            if (med.getGraphOptions().getDisplayMode() ==
                        GraphOptionsInterface.CROSSHAIR_MODE)
                setCursor(Cursor.getDefaultCursor());
            if (med.getGraphOptions().getDisplayMode() ==
                        GraphOptionsInterface.ZOOM_MODE)
                setCursor(Cursor.getPredefinedCursor(
                            Cursor.CROSSHAIR_CURSOR));
            if (med.getGraphOptions().getDisplayMode() ==
                        GraphOptionsInterface.GRAB_MODE)
                setCursor(Cursor.getPredefinedCursor(
                            Cursor.HAND_CURSOR));
        }
        public void mouseClicked(MouseEvent e)
        {
            if (e.getClickCount()>1) //If double click
            {
                //Reset graph to draw complete waveform
                med.getZoomDecorator().setZooming(false);
                med.updateData(med.getGraphData());
                GraphicsFacade.this.repaint();
```

```java
        }
    }
    public void mouseEntered(MouseEvent e) {}
    public void mouseExited(MouseEvent e) {}
    public void mouseDragged(MouseEvent e)
    {
        int currX = e.getX();
        int currY = e.getY();
        //If the current display mode is GRAB,
        //we need to change med,
        //so that the waveform is dragged
        //For the other modes (zoom, crosshair) no
        //change in the waveform
        //is required during mouse drag
        switch (med.getGraphOptions().getDisplayMode())
        {
            case GraphOptionsInterface.GRAB_MODE:
                if (prevX == currX && prevY == currY)
                    return; //Mouse not dragged

                //Change waveform limits in med to
                //perform dragging
                double deltaXData = med.getXValue(prevX)
                            - med.getXValue(currX);
                double deltaYData = med.getYValue(prevY)
                            - med.getYValue(currY);;
                med.setMaxXData(med.getMaxXData()
                                        + deltaXData);
                med.setMinXData(med.getMinXData()
                                        + deltaXData);
                med.setMaxYData(med.getMaxYData()
                                        + deltaYData);
                med.setMinYData(med.getMinYData()
                                        + deltaYData);
                break;
            case GraphOptionsInterface.CROSSHAIR_MODE:
                med.getCrossHairDecorator().
                setCurrX(currX);
                break;
            case GraphOptionsInterface.ZOOM_MODE:
                med.getZoomDecorator().setEndX(currX);
                med.getZoomDecorator().setEndY(currY);
                break;
        }
        prevX = currX;
        prevY = currY;

        //Force repaint
        GraphicsFacade.this.repaint();
```

```
}
public void mouseReleased(MouseEvent e)
{
    //Mouse released actions are only required
    //when in ZOOM mode
    //In this case, med parameters are changed so
    //that the waveform
    //region fits the zoom box
    if (med.getGraphOptions().getDisplayMode() ==
        GraphOptionsInterface.ZOOM_MODE)
    {
        int currX = e.getX();
        int currY = e.getY();
        if (startX == currX && startY == currY)
            return; //Not dragged
        //Set minimum width and height for the
        //zoom box
        if (startX == currX) currX = startX + 1;
        if (startY == currY) currY = startY + 1;
        //Get the values corresponding to the
        //limits of the zoom box
        //Need to handle every direction in which
        //mouse has been moved
        double minXData, minYData, maxXData,
               maxYData;
        if (currX > startX)
        {
            minXData = med.getXValue(startX);
            maxXData = med.getXValue(currX);
        }
        else
        {
            minXData = med.getXValue(currX);
            maxXData = med.getXValue(startX);
        }
        if (currY < startY)
        {
            minYData = med.getYValue(startY);
            maxYData = med.getYValue(currY);
        }
        else
        {
            minYData = med.getYValue(currY);
            maxYData = med.getYValue(startY);
        }
        //Let the values corresponding to the
        //limits of the zoom box
        //be the limits of the displayed waveform
        med.setMaxXData(maxXData);
```

```
                med.setMaxYData(maxYData);
                med.setMinXData(minXData);
                med.setMinYData(minYData);
                med.getZoomDecorator().setZooming(false);
                //Force repaint
                GraphicsFacade.this.repaint();
            }
        }
    }
}
```

E.6 The GUI Domain

The GUI domain comprises six classes. There are three dialogs:

- `ConnectDialog`
- `GraphOptionsDialog`
- `OpenDialog`

together with a class which describes the colour panel of the graph options dialog:

- `ColorItemPanel`

There is the class which holds the graph options data:

- `GraphOptions`

and there is the GUI facade class:

- `GuiFacade`

E.6.1 The Dialog Classes

`GraphOptionsDialog`

The `GraphOptionsDialog` class reads:

```
package guiDomain;

import sharedDataInterfaces.*;
import javax.swing.*;
import java.awt.*;
import java.awt.event.*;
import java.util.*;
import javax.swing.border.*;
import java.lang.*;
/**   A dialog frame which allows a user to
      alter a GraphOptions object*/
public class GraphOptionsDialog
    extends JDialog
```

```java
{
    private GraphOptions options;
    private GuiFacade parentGui;
    private ColorItemPanel titlePanel;
    private ColorItemPanel graphPanel;
    private ColorItemPanel linePanel;
    // border options
    private JRadioButton defaultBorderButton =
                            new JRadioButton("Default");
    private JRadioButton zeroBorderButton =
                            new JRadioButton("Zero");
    // tick number options
    private JRadioButton flexibleNumberButton =
                            new JRadioButton("Flexible");
    private JRadioButton niceNumberButton =
                            new JRadioButton("Nice");
    // display mode buttons
    private JRadioButton zoomButton =
                            new JRadioButton("Zoom");
    private JRadioButton crossHairButton =
                            new JRadioButton("CrossHair");
    private JRadioButton grabButton =
                            new JRadioButton("Grab");
    /** The constructor receives handles
        to the owner frame
        and the GraphOptions object*/
    public GraphOptionsDialog(JFrame owner, GraphOptions go)
    {
        super(owner, "eScope Graph Options", true);
        parentGui = (GuiFacade) owner;
        setBounds(10, 100, 350, 400);

        JPanel mainPanel = new JPanel();
        JPanel subPanel = new JPanel();
        JPanel colorPanel = new JPanel();
        JPanel plotPanel = new JPanel();
        JPanel mousePanel = new JPanel();
        JPanel buttonPanel = new JPanel();

        options = go;

        mainPanel.setLayout(new BorderLayout());
        subPanel.setLayout(new GridLayout(3, 1));
        colorPanel.setLayout(new GridLayout(3, 1));
        plotPanel.setLayout(new GridLayout(2, 1));
        mousePanel.setLayout(new GridLayout(2, 1));
        buttonPanel.setLayout(new BorderLayout());
        titlePanel = new ColorItemPanel("Graph Title",
                                GraphOptions.colorNames);
```

```java
            graphPanel = new ColorItemPanel("Background",
                                GraphOptions.colorNames);
            linePanel = new ColorItemPanel("Plot Line",
                                GraphOptions.colorNames);
    colorPanel.add(titlePanel);
    colorPanel.add(graphPanel);
    colorPanel.add(linePanel);
    setOptions();
    colorPanel.setBorder(new LineBorder(Color.black));
    //plotPanel
    JPanel borderPanel = new JPanel();
    borderPanel.setLayout(new FlowLayout());
    borderPanel.add(new JLabel("Graph Border"));
    ButtonGroup bg = new ButtonGroup();
    bg.add(defaultBorderButton);
    bg.add(zeroBorderButton);
    borderPanel.add(defaultBorderButton);
    borderPanel.add(zeroBorderButton);
    plotPanel.add(borderPanel);

    JPanel tickNumberPanel = new JPanel();
    tickNumberPanel.setLayout(new FlowLayout());
    tickNumberPanel.add(new JLabel("Tick Numbers"));
    bg = new ButtonGroup();
    bg.add(niceNumberButton);
    bg.add(flexibleNumberButton);
    tickNumberPanel.add(niceNumberButton);
    tickNumberPanel.add(flexibleNumberButton);
    plotPanel.add(tickNumberPanel);
    plotPanel.setBorder(new LineBorder(Color.black));

    // mouse options panel
    JPanel displayModePanel = new JPanel();
    displayModePanel.setLayout(new FlowLayout());
    displayModePanel.add(new JLabel("Mouse mode: "));
    bg = new ButtonGroup();
    bg.add(crossHairButton);
    bg.add(zoomButton);
    bg.add(grabButton);
    displayModePanel.add(crossHairButton);
    displayModePanel.add(zoomButton);
    displayModePanel.add(grabButton);
    mousePanel.add(displayModePanel);

    JPanel infoPanel = new JPanel();
    JLabel mouseInfo = new JLabel(
            "(Double click for default view)");
    infoPanel.add(mouseInfo);
    mousePanel.add(infoPanel);
```

E.6 The GUI Domain

```java
        mousePanel.setBorder(new LineBorder(Color.black));

        // button panel
        JButton okButton = new JButton(new OKAction("OK"));
        JButton applyButton = new JButton(
                                    new ApplyAction("Apply"));
        JButton cancelButton = new JButton(
                                    new CancelAction("Cancel"));
        JPanel buttons = new JPanel();
        buttons.add(okButton);
        buttons.add(applyButton);
        buttons.add(cancelButton);
        buttonPanel.add(buttons,"South");
        buttonPanel.setBorder(new LineBorder(Color.black));
        subPanel.add(colorPanel);
        subPanel.add(plotPanel);
        subPanel.add(mousePanel);

        mainPanel.add(subPanel,"North");
        mainPanel.add(buttonPanel,"South");
        getContentPane().add(mainPanel);
        setModal(false);
        pack();
    }

    /** Updates the GUI display after options selected */
    public void setOptions()
    {
        titlePanel.setSelected(options.getTitleColor());
        graphPanel.setSelected(options.getGraphColor());
        linePanel.setSelected(options.getLineColor());
        if (options.getDefaultBorder())
        {
            defaultBorderButton.setSelected(true);
        }
        else
        {
            zeroBorderButton.setSelected(true);
        }
        if (options.getNiceNumbers())
        {
            niceNumberButton.setSelected(true);
        }
        else
        {
            flexibleNumberButton.setSelected(true);
        }
        switch(options.getDisplayMode())
        {
```

```java
                    case GraphOptionsInterface.ZOOM_MODE:
                        zoomButton.setSelected(true);
                        break;
                    case GraphOptionsInterface.CROSSHAIR_MODE:
                        crossHairButton.setSelected(true);
                        break;
                    case GraphOptionsInterface.GRAB_MODE:
                        grabButton.setSelected(true);
                        break;
            }
    }
    /** Updates the GraphOptions object
        following user selection */
    public void updateOptions()
    {
        options.setTitleColor(titlePanel.getSelected());
        options.setGraphColor(graphPanel.getSelected());
        options.setLineColor(linePanel.getSelected());
        if (defaultBorderButton.isSelected())
        {
            options.setDefaultBorder(true);
        }
        else
        {
            options.setDefaultBorder(false);
        }
        if (niceNumberButton.isSelected())
        {
            options.setNiceNumbers(true);
        }
        else
        {
            options.setNiceNumbers(false);
        }
        if(zoomButton.isSelected())
            options.setDisplayMode(
                GraphOptionsInterface.ZOOM_MODE);
        if(crossHairButton.isSelected())
            options.setDisplayMode(
                GraphOptionsInterface.CROSSHAIR_MODE);
        if(grabButton.isSelected())
            options.setDisplayMode(
                GraphOptionsInterface.GRAB_MODE);
    }
    /** Inner class which handles the OK button action */
    private class OKAction
        extends AbstractAction
    {
        public OKAction(String s)
```

```
                {
                    super(s);
                }
                public void actionPerformed(ActionEvent event)
                {
                    updateOptions();
                    setVisible(false);
                }
            }
        }
        /**    Inner class which handles the Apply button
               action */
        private class ApplyAction
            extends AbstractAction
        {
            public ApplyAction(String s)
            {
                super(s);
            }
            public void actionPerformed(ActionEvent event)
            {
                updateOptions();
                parentGui.applyGraphOptions();
            }
        }
        /**    Inner class which handles the Cancel button
               action */
        private class CancelAction
            extends AbstractAction
        {
            public CancelAction(String s)
            {
                super(s);
            }

            public void actionPerformed(ActionEvent event)
            {
                setVisible(false);
            }
        }
    }
```

ColorItemPanel

The `ColorItemPanel` class reads:

```
package guiDomain;

import javax.swing.*;
```

```java
import java.awt.*;
public class ColorItemPanel extends JPanel
{
    private JComboBox colorBox;
    /** The names of colors that appear in the combo box */
    private String[] colorNames;
    public ColorItemPanel(String name, String[] cNames)
    {
        colorNames = cNames;
        setLayout(new GridLayout(1,2));
        colorBox = new JComboBox(colorNames);
        add(new JLabel(name, SwingConstants.CENTER));
        add(colorBox);
    }
    /**
            Sets the selected item in the combo box
    */
    public void setSelected(int i)
    {
            colorBox.setSelectedIndex(i);
    }
    /**
            Gets the selected item in the combo box
    */
    public int getSelected()
    {
            return colorBox.getSelectedIndex();
    }
}
```

E.6.2 GraphOptions

The GraphOptions class reads:

```java
package guiDomain;
import sharedDataInterfaces.GraphOptionsInterface;

import javax.swing.*;
import java.awt.*;
import java.awt.event.*;
import java.util.*;
import javax.swing.border.*;
/** Stores the options settings for the graph panel */
public class GraphOptions implements GraphOptionsInterface
{
    // Color indices for Colors
    private int titleColor=RED, graphColor=WHITE,
                lineColor=BLACK;
```

```java
    private boolean defaultBorder=true;
    private boolean niceNumbers=true;
    private int displayMode = CROSSHAIR_MODE;
    public GraphOptions() {}

    public Color selectColor(int i)
    {
        switch (i)
        {
            case 0:
                return Color.black;
            case 1:
                return Color.white;
            case 2:
                return Color.red.darker();
            case 3:
                return Color.blue;
            case 4:
                return Color.orange;
            case 5:
                return Color.gray;
            case 6:
                return Color.lightGray;
            case 7:
                return Color.darkGray;
            case 8:
                return Color.magenta;
            case 9:
                return Color.green;
            case 10:
                return Color.yellow;
            case 11:
                return Color.cyan;
            case 12:
                return Color.pink;
        }
        return Color.black; //index out of range
    }
    public String selectColorName(int i)
    {
            if (i>=0 && i<numColors)
                return colorNames[i];
            return ""; // index out of range
    }
    public void setTitleColor(int colorIndex)
                        {titleColor=colorIndex;}
    public int getTitleColor() {return titleColor;}
    public void setGraphColor(int colorIndex)
                        {graphColor=colorIndex;}
```

```
        public int getGraphColor() {return graphColor;}
        public void setLineColor(int colorIndex)
                                {lineColor=colorIndex;}
        public int getLineColor() {return lineColor;}
        public void setDefaultBorder(boolean isDefault)
                                {defaultBorder=isDefault;}
        public boolean getDefaultBorder()
                                {return defaultBorder;}
        public void setNiceNumbers(boolean isNice)
                                {niceNumbers=isNice;}
        public boolean getNiceNumbers() {return niceNumbers;}
        public void setDisplayMode(int mode)
                                {displayMode = mode;}
        public int getDisplayMode() {return displayMode;}
}
```

E.6.3 GuiFacade

The GuiFacade class reads:

```
package guiDomain;
import sharedInterfaces.*;
import sharedDataInterfaces.*;

import java.io.*;
import java.awt.*;
import java.awt.event.*;
import javax.swing.*;
import javax.swing.event.*;
import javax.swing.tree.*;
import java.awt.geom.*;
import java.awt.font.*;
import java.text.*;
import java.util.*;

public class GuiFacade extends JFrame
                    implements GuiFacadeInterface
{   // Handles on other domains
    private DataServerFacadeInterface database;
    private AbstractGraphicsFacade graph;
    private GraphOptions options = new GraphOptions();

    // Components for the menu
    private JMenuBar menuBar;

    private JMenu fileMenu;
    private JMenuItem fileConnectItem;
    private JMenuItem fileOpenItem;
```

```java
private JMenuItem fileExitItem;

private JMenu graphMenu;
private JMenuItem graphOptionsItem;

// Components for the main window
private JLabel description;

// Dialogs
private OpenDialog openD;
private ConnectDialog connectD;
private GraphOptionsDialog graphD;

// file tree
private JTree tree;

public GuiFacade()
{
    // Construct the menu
    menuBar = new JMenuBar();
    setJMenuBar( menuBar );
    fileMenu = new JMenu( "File" );
    menuBar.add( fileMenu );
    fileConnectItem = new JMenuItem(
                            new ConnectAction() );
    fileMenu.add( fileConnectItem );
    fileOpenItem = new JMenuItem( new OpenAction() );
    fileMenu.add( fileOpenItem );
    fileMenu.addSeparator();
    fileExitItem = new JMenuItem( new ExitAction() );
    fileMenu.add( fileExitItem );
    // Graph Menu
    graphMenu = new JMenu( "Graph" );
    menuBar.add( graphMenu );
    graphOptionsItem = graphMenu.add(
                            new GraphOptionsAction() );
    description = new JLabel();
    connectD = new ConnectDialog(
            this, "Enter Connection Details", true);
    openD = new OpenDialog(
            this, "Enter experiment and shot", true);
    graphD = new GraphOptionsDialog(this, options);
}

public void initialise(DataServerFacadeInterface _source,
                    AbstractGraphicsFacade _graph)
{
    database = _source;
    graph = _graph;
```

```
            graph.setGraphOptions(options);
            tree=new JTree(new DefaultTreeModel(
                    new DefaultMutableTreeNode
                    ("No experiemnt open")));
            tree.addTreeSelectionListener(
                        new DataTreeSelectionListener() );
            JScrollPane treeScroll = new JScrollPane( tree );
            treeScroll.setMinimumSize(
                        new Dimension( 150, 10 ) );

            JSplitPane rightPane = new JSplitPane(
                        JSplitPane.VERTICAL_SPLIT,
                        new JScrollPane( description ),graph);
            JSplitPane mainPane = new JSplitPane(
                        JSplitPane.HORIZONTAL_SPLIT,
                        treeScroll, rightPane );
            getContentPane().add( mainPane, "Center" );
            this.setDefaultCloseOperation(JFrame.EXIT_ON_CLOSE);
            this.setTitle( "No experiment opened" );
            this.setBounds( 200, 200, 800, 400 );
            this.show();
      }
      public void graphPointUpdated(double x,
                                double y, boolean inRange)
      {
            if (inRange)
            {
                DecimalFormat decForm =
                            new DecimalFormat("0.##E0" );
                String xString = decForm.format(x);
                String yString = decForm.format(y);
                description.setText(" x= " + xString +
                                    "   y= " + yString );
            }
            else description.setText(" ");
      }
      public void setDescription(String text)
      {
            description.setText(text);
      }
      public void applyGraphOptions()
      {
            graph.applyGraphOptions();
            graph.repaint();
      }
      // Private Inner Classes
      private class ExitAction extends AbstractAction
      {
            public ExitAction() {   super( "Exit" );}
```

```java
        public void actionPerformed( ActionEvent event )
        {    System.exit( 0 );
        }
    }
    private class ConnectAction extends AbstractAction
    {
        public ConnectAction(){    super( "Connect ..." );}

        public void actionPerformed( ActionEvent event )
        {
            if (connectD.showDialog())
            {
                String server = connectD.getServer();
                try
                {
                    if ( database.isConnected() )
                    {
                        database.disconnect();
                    }
                    database.connect( server );
                    String newTitle =
                    "Connection made. No experiment opened";
                    GuiFacade.this.setTitle(newTitle);
                }
                catch (IOException e)
                {
                    JOptionPane.showMessageDialog(
                        GuiFacade.this,
                        "Failed to connect to server",
                        "Error",
                        JOptionPane.ERROR_MESSAGE );
                }
            }
        }
    }
    private class OpenAction extends AbstractAction
    {
        public OpenAction(){ super( "Open ..." );}

        public void actionPerformed( ActionEvent event )
        {    if (openD.showDialog())
            {
                String experiment = openD.getExperiment();
                int shot = openD.getShot();
                try
                {
                    if ( database.isOpen() )
                    {
```

```java
                        database.close();
                    }
                    database.open( experiment, shot );
                    database.constructTree(tree);
                    String newTitle =
                        " Opened experiment: "
                        + experiment + " shot: " + shot;
                    GuiFacade.this.setTitle(newTitle);
                }
                catch (IOException e)
                {
                    JOptionPane.showMessageDialog
                        (GuiFacade.this,
                        "Failed to open " +
                        "experiment " +
                        experiment +", shot " +shot + ": " +
                        e.getMessage(), "Error",
                        JOptionPane.ERROR_MESSAGE);
                }
            }
        }
    }
}
/** Open the graph options dialog */
private class GraphOptionsAction extends AbstractAction
{
    public GraphOptionsAction(){ super("Options" );}

    public void actionPerformed(ActionEvent event)
    {
        if (graphD == null)
            graphD = new GraphOptionsDialog
                    (GuiFacade.this, options);
        graphD.setOptions();
        graphD.show();
        GuiFacade.this.applyGraphOptions();
    }
}
/** Plots the graph and displays the
    path of the selected tree node */
private class DataTreeSelectionListener
        implements TreeSelectionListener
{
    public void valueChanged( TreeSelectionEvent e )
    {    // Get new data and plot it
        Object path[];
        path = e.getPath().getPath();
        String shortFileAddress="";
        for ( int i = 1; i < path.length; i ++ )
        {
```

```
                    shortFileAddress = shortFileAddress
                                     + path[i].toString();
                }
                if (!shortFileAddress.equals(""))
                    // (Do not plot or display the "TOP" node)
                {
                    GuiFacade.this.setDescription(
                                        shortFileAddress);
                    GuiFacade.this.graph.setGraphData
                        (database.getPlotData(shortFileAddress));
                    GuiFacade.this.graph.repaint();
                        //(prints a message if no data)
                }
            }
        }
    }
}
```

F
Excerpts from Later Listings

F.1 EScope5

F.1.1 GraphMediator

The following excerpt shows the changes to the decorator chaining and the **recordClosestWave** method.

```
public void createDecorator()
{
    /*Build the decorator chain. The elements in the
      chain define what is going to be displayed.

        If no data or bad data then the decorator chain is
        only composed of a DrawCentredMessage object
        containing an error message.*/
    if (graphData == null || !isGraphData)
    {
        if (graphData != null && isDataError)
            warningDecorator.setMessage(
                "Unexpected data. See console message.");
        else
            warningDecorator.setMessage("No data");

        //Build decorator chain with a centred message
        decorator = warningDecorator;
    }
    else //Build the decorator chain based
         //on the current display mode
    {
        if (graphOptions.getDisplayMode()
                        ==graphOptions.getCROSSHAIR())
            //decorator chain with X and Y axis,
            //crosshair
            decorator = concatenateDecorators(
```

```java
                        labelsDecorator,
                        crossHairDecorator,
                        axisYDecorator,
                        axisXDecorator);
        if (graphOptions.getDisplayMode()
                        ==graphOptions.getZOOM())
            //decorator chain with X and Y axis,
            //zoom box
            decorator = concatenateDecorators(
                        labelsDecorator,
                        zoomDecorator,
                        axisYDecorator,
                        axisXDecorator);
        if (graphOptions.getDisplayMode()
                        ==graphOptions.getGRAB())
            //decorator chain with X and Y axis
            decorator = concatenateDecorators(
                        labelsDecorator,
                        axisYDecorator,
                        axisXDecorator);
        //Add waveforms
        for (int nWave = 0; nWave<graphData.length;
                nWave++)
        {
            DrawWaveform wave = new DrawWaveform(
                                        this,nWave);
            decorator = concatenateDecorators(
                                        wave,decorator);
        }
    }
}
/** Forms decorator chain. Note variable argument
    list. */
private GraphDecorator
        concatenateDecorators(
                        GraphDecorator ... decorators)
{
    for (int i = decorators.length - 2; i >= 0; i--)
        decorators[i].setDecorator(decorators[i + 1]);
    return decorators[0];
}
public GraphDecorator getDecorator()
{
    return decorator; // move to accessor methods?
}
//For Cross-Hair
//Find the closest wave to the point (currX, currY)
public void recordClosestWave()
{
```

```
        double xVal = getXValue(currX);
        double yVal = getYValue(currY);
        double minDist =
                    graphData[0].getYDistance(xVal,yVal);
        closestGraphIdx = 0;
        for(int i = 1; i < graphData.length; i++)
        {
            double currDist =
                        graphData[i].getYDistance(xVal,yVal);
            if(currDist < minDist)
            {
                minDist = currDist;
                closestGraphIdx = i;
            }
        }
    }
```

F.1.2 GraphDataInGraphics

The following excerpt shows the additional methods needed to find the closest waveform from a mouse click.

```
    /**Find an index which corresponds to the data value 'x'*/
    private int getXIdx(double x)
    {
        double[] xArray = graphData.getX();
        int xIdx = xArray.length / 2;
        int minIdx = 0, maxIdx = xArray.length - 1;
        boolean found = false;

        //Bisect the interval in O(log n) operations
        while (!found)
        {
            if (xIdx == 0 || xIdx == xArray.length - 1 ||
                (xArray[xIdx] >= x && xArray[xIdx - 1] <= x) ||
                (xArray[xIdx] <= x && xArray[xIdx + 1] >= x) ||
                xIdx == maxIdx || xIdx == minIdx)
                found = true;
            else
            {
                if (xArray[xIdx] > x)
                    maxIdx = xIdx;
                else
                    minIdx = xIdx;
                xIdx = (maxIdx + minIdx) / 2;
            }
        }
        return xIdx;
    }
```

```java
/** return "vertical", y, distance corresponding to x*/
public double getYDistance(double x, double y)
{
    double[] yArray = graphData.getY();
    int xIdx = getXIdx(x);

    return Math.abs(yArray[xIdx] - y);
}
```

F.2 EScope6

F.2.1 Reading Properties from GuiFacade

```java
//Read from property file 'EScope.properties'
//The number of columns is defined
//by the property label 'num_columns'
//For column i, the number of rows is defined
//by the property label
// 'column_<i>.num_rows'
private void getRowColumns()
{
    Properties props = new Properties();
    try {
        props.load(new FileInputStream
                            ("EScope.properties"));
        numCols = Integer.parseInt(props.
                        getProperty("num_columns"));
        rowArray = new int[numCols];
        for(int i = 0; i < numCols; i++)
        {
            rowArray[i] = Integer.parseInt(
                    props.getProperty
                    ("column_"+(i+1)+".num_rows"));
        }
    }catch(Exception exc)
    //If anything goes wrong (e.g. property file missing)
    //default to 2 columns, each with two rows
    {
        System.out.println("Exception in " +
            "MultipleGraphicsFacade.getRowColums: " + exc);
        numCols = 2;
        rowArray = new int[]{2,2};
    }
}
```

F.2.2 GraphicsFacade

Declarations and Construction

```java
package graphicsDomain;

import sharedInterfaces.*;
import sharedDataInterfaces.*;

import java.awt.*;
import java.awt.event.*;
import java.awt.geom.*;
import java.awt.font.*;
import javax.swing.*;
import java.awt.image.*;
import java.util.*;

public class GraphicsFacade extends AbstractGraphicsFacade
{
    private GraphMediator med;

    private GraphMouseHandler mouseHandler =
                            new GraphMouseHandler();
    private GuiFacadeInterface guiFacade;
                //Used to draw current waveform coordinates
    //The popup menu used to select the window
    //and to globally set scales
    private JPopupMenu popupMenu;

    //Current mouse state
    private static final int RELEASED = 1,
        JUST_PRESSED = 2, JUST_RELEASED = 3,
        DRAGGING = 4;
    private int mouseState = RELEASED;

    // The following instances are used to
    //draw into the image buffer
    private BufferedImage wavesImage;
            //Offline waveform image
    //Current Translation for waves image to be copied in
    //repaint
    private AffineTransform currWaveTrans =
                            new AffineTransform();
    //Base Transform, computed every time the
    // offline image is created
    private AffineTransform baseWaveTrans;
    //Previous window X and Y size.
    private int prevXSize = -1, prevYSize = -1;
    private boolean forceImageUpdate=false;
```

```
public GraphicsFacade()
{
    med = new GraphMediator(this);
    addMouseListener(mouseHandler);
    addMouseMotionListener(mouseHandler);
    createPopupMenu();
}

public AbstractGraphicsFacade getGraphicsFacade()
{
    GraphicsFacade newGraph = new GraphicsFacade();
    return newGraph;
}
```

Excerpt from `paintComponent`

```
public void paintComponent(Graphics g)
{
    ....
    //call draw method of the decorator chain
    GraphDecorator decorator = med.getDecorator();
    GraphDecorator waveDecorator =
                    med.getWaveDecorator();

    if (waveDecorator==null)
    {
            decorator.draw(g2);
    }
    else
    {
        //Check if window size, or waveform limits,
        //changed, or mouse
        //just pressed or released.
        //In these cases, waveforms need to be
        //plotted again in the
        //image buffer.
        if (windowChanged() || mouseActive() ||
            forceImageUpdate)
        {
            updateWavesImage(waveDecorator);
            prevXSize = xSize;
            prevYSize = ySize;
            forceImageUpdate=false;
        }
        //Copy image buffer onto this panel
        if (wavesImage != null)
            g2.drawImage(wavesImage,
                        currWaveTrans, this);
```

```
            //Now draw cursors, labels and axes
            decorator.draw(g2);
        }

        //If this window is selected,
        //draw a red rectangle around it
        if (multiGraph.getSelectedGraph() == this)
        {
            g2.setColor(Color.RED);
            g2.setStroke(med.getGraphOptions().
                        getLineStroke());
            g2.drawRect(0, 0, xSize - 1, ySize - 1);
        }
        //adjust mouse state
        switch (mouseState)
        {
            case JUST_PRESSED:
                mouseState = DRAGGING;
                break;
            case JUST_RELEASED:
                mouseState = RELEASED;
        }
    }
```

Convenience Methods to Distinguish Repaints

```
    //Tests whether a window has changed size.
    private boolean windowChanged()
    {
        return (prevXSize != med.getXSize()
            || prevYSize != med.getYSize());
    }

    //Tests whether a mouse interaction has started or
    //finished
    private boolean mouseActive()
    {
        return (mouseState == JUST_RELEASED ||
                mouseState == JUST_PRESSED);
    }
```

Drawing Waveforms onto Image Buffer

```
    /** Draw waveforms in the image buffer.*/
    private void updateWavesImage(
                    GraphDecorator waveDecorator)
    {
        // Make sure that this image is large enough to
        // accomodate waveforms when dragging.
        // (Note that very large buffers waste heap space)
```

```
        int fullWidth = med.getXSize() * 3;
        int fullHeight = med.getYSize() * 3;

        //Create the offline image
        //only if the image previously created is smaller
        wavesImage = (BufferedImage)
                        createImage(fullWidth, fullHeight);
        Graphics2D wavesGraphics =
                        wavesImage.createGraphics();

        //Start drawing waveforms in buffered image
        wavesGraphics.setColor(med.getGraphOptions().
                    selectColor(
                    med.getGraphOptions().getGraphColor()));
        wavesGraphics.fill(new Rectangle2D.Double(0, 0,
                                fullWidth, fullHeight));

        //Adjust proper translation when drawing so that
        //entire waveform set fits in the buffered image
        int deltaX = med.getXSize();
        int deltaY = med.getYSize();

        currWaveTrans.setToTranslation(deltaX, deltaY);
        wavesGraphics.transform(currWaveTrans);
        //Draw waves in buffered image
        waveDecorator.draw(wavesGraphics);
        //Adjust now translation to the opposite way, so that
        //waveforms will be correctly displayed on the screen
        currWaveTrans.setToTranslation(-deltaX, -deltaY);
        //current translation will be
        //updated when dragging mouse in GRAB mode
        baseWaveTrans = new AffineTransform(currWaveTrans);
                //Record this tranformation
    }
```

Pop-up Menu and `selectWindow` Method

```
    /** Creates popup dialog for use with right mouse
        button */
    private void createPopupMenu()
    {
        popupMenu = new JPopupMenu();
        JMenuItem selectI = new JMenuItem("Select");
        selectI.addActionListener(new ActionListener()
        {
            public void actionPerformed(ActionEvent e)
            {
                selectWindow();
            }
```

```java
        });
        popupMenu.add(selectI);
        popupMenu.pack();
        popupMenu.setInvoker(this);
    }

    /** Select this facade window,
        possibly de-selecting another one */
    private void selectWindow()
    {
        multiGraph.setSelectedGraph(GraphicsFacade.this);
        //Force repaint of the parent panel,
        //so that the GraphicsFacade
        //which looses focus can redraw itself
        GraphicsFacade.this.getParent().repaint();
    }
```

mouseDragged Method Showing Image Translation

```java
        public void mouseDragged(MouseEvent e)
        {
            int currX = e.getX();
            int currY = e.getY();
            mouseState = DRAGGING;

            //If the current display mode is GRAB,
            //we need to change med,
            //so that the waveform is dragged
            //For the other modes (zoom, crosshair) no
            //change in the waveform
            //is required during mouse drag
            switch (med.getGraphOptions().getDisplayMode())
            {
                case GraphOptionsInterface.GRAB_MODE:
                    if (prevX == currX && prevY == currY)
                        return; //Mouse not dragged

                    //Change waveform limits in med
                    //to perform dragging
                    double deltaXData = med.getXValue(prevX) -
                                        med.getXValue(currX);
                    double deltaYData = med.getYValue(prevY) -
                                        med.getYValue(currY);;
                    med.setMaxXData(med.getMaxXData()
                                        + deltaXData);
                    med.setMinXData(med.getMinXData()
                                        + deltaXData);
                    med.setMaxYData(med.getMaxYData()
                                        + deltaYData);
```

```
                        med.setMinYData(med.getMinYData()
                                                    + deltaYData);
                        //The waveform image needs to be
                        //translated as well
                        currWaveTrans.setToTranslation(
                                currWaveTrans.getTranslateX()
                                                    + currX − prevX,
                                currWaveTrans.getTranslateY()
                                                    + currY − prevY);
                        break;
                    case GraphOptionsInterface.CROSSHAIR_MODE:
                        med.setCurrX(currX);
                        break;
                    case GraphOptionsInterface.ZOOM_MODE:
                        med.getZoomDecorator().setEndX(currX);
                        med.getZoomDecorator().setEndY(currY);
                        break;
                }
                prevX = currX;
                prevY = currY;

                //Force repaint
                GraphicsFacade.this.repaint();
        }
```

F.3 EScope7

F.3.1 GraphUpdateEvent

```
package graphicsDomain;

class GraphUpdateEvent
{
    double minX, maxX, minY, maxY;

    //Constructors
    GraphUpdateEvent(double minX, double maxX)
    {
        this.minX = minX;
        this.maxX = maxX;
    }
    GraphUpdateEvent(double minX, double maxX,
                                double minY, double maxY)
    {
        this.minX = minX;
        this.maxX = maxX;
        this.minY = minY;
        this.maxY = maxY;
    }
```

```java
    //Accessor methods
    public double getMinX() {return minX;}
    public double getMaxX() {return maxX;}
    public double getMinY() {return minY;}
    public double getMaxY() {return maxY;}
}
```

F.3.2 The Scale Interfaces

```java
package graphicsDomain;

public interface GraphScaleManager
{
    public void addGraphScaleListener(GraphScaleListener l);

    public void removeGraphScaleListener(
                                GraphScaleListener l);

    public void notifyGraphScaleUpdated(GraphUpdateEvent e);

    public void notifyGraphXScaleUpdated(GraphUpdateEvent e);
}
```

```java
package graphicsDomain;

public interface GraphScaleListener
{
    public void graphScaleUpdated(GraphUpdateEvent e);

    public void graphXScaleUpdated(GraphUpdateEvent e);
}
```

F.3.3 GraphicsFacade: Pop-up Menu and Associated Methods

```java
    /** Creates popup dialog. Right click on any graph
        window. */
    private void createPopupMenu()
    {
        popupMenu = new JPopupMenu();
        JMenuItem selectI = new JMenuItem("Select");
        selectI.addActionListener(new ActionListener()
        {
            public void actionPerformed(ActionEvent e)
            {
                selectWindow();
            }
        });
        popupMenu.add(selectI);
```

```java
            JMenuItem allSameXScaleI = new JMenuItem(
                                        "All same X scale");
            allSameXScaleI.addActionListener(new ActionListener()
            {
                public void actionPerformed(ActionEvent e)
                {
                    allSameXScale();
                }
            });
            popupMenu.add(allSameXScaleI);
            JMenuItem allSameScaleI = new JMenuItem(
                                        "All same scale");
            allSameScaleI.addActionListener(new ActionListener()
            {
                public void actionPerformed(ActionEvent e)
                {
                    allSameScale();
                }
            });
            popupMenu.add(allSameScaleI);
            popupMenu.pack();
            popupMenu.setInvoker(this);
    }

    /** Select this graph window,
        possibly de-selecting another one */
    private void selectWindow()
    {
        guiFacade.setSelectedGraph(GraphicsFacade.this);
        //Force repaint of the parent panel,
        //so that the GraphicsFacade
        //which looses focus can redraw itself
        GraphicsFacade.this.getParent().repaint();
    }

    /** Set the same X scale for all windows */
    private void allSameXScale()
    {
        GraphUpdateEvent graphUpdateEvent =
            new GraphUpdateEvent(
                med.getMinXData(), med.getMaxXData());
        notifyGraphXScaleUpdated(graphUpdateEvent);
    }

    /** set the same scale for all graph windows */
    private void allSameScale()
    {
        GraphUpdateEvent graphUpdateEvent =
            new GraphUpdateEvent(
```

```
            med.getMinXData(), med.getMaxXData(),
            med.getMinYData(), med.getMaxYData());
    notifyGraphScaleUpdated(graphUpdateEvent);
}
```

F.4 EScope8

F.4.1 DataServerProxy: getPlotData

```
/**
 Gets the plot data (x-values, y-values,
 x-units, y-units) for a node.
 Once loaded saved locally in tmp file.
 On request for same node, local data is used.
 @param inPath Specifies the path of the node to be
  retrieved
 @return a GraphData object which stores waveform data
 */
public GraphDataInterface getPlotData(String inPath)
{
    GraphDataInterface output = null;

    try
    {
        // First check whether graph is already saved
        // locally
        for (int i = 0; i < graphDataList.size(); i++)
        {
            if (inPath.equals((String)
                              graphDataList.get(i)))
            {
                FileInputStream fis =
                        new FileInputStream(i + ".tmp");
                ObjectInputStream ois =
                        new ObjectInputStream(fis);
                output =
                    (GraphDataInterface) ois.readObject();
                ois.close();
                fis.close();

                System.out.println("Graph " +
                        inPath + " is loaded from " +
                            i + ".tmp");
                return output;
            }
        }

        // Not locally saved, so get it from real
        // retriever
```

```java
                output = dataServer.getPlotData(inPath);

                // Now save it locally for next time
                FileOutputStream fos = new FileOutputStream(
                                graphDataList.size() + ".tmp");
                ObjectOutputStream oos =
                                new ObjectOutputStream(fos);
                oos.writeObject(output);
                oos.close();
                fos.close();
                System.out.println("Graph " + inPath +
                                " is stored in " +
                                graphDataList.size() + ".tmp");
                graphDataList.add(inPath);
        }
```

F.5 EScope10

F.5.1 ServerSelectDialog

```java
package guiDomain;

import java.awt.*;
import java.awt.event.*;
import javax.swing.*;
import java.util.*;
import java.io.*;

public class ServerSelectDialog extends JDialog
{
    private JButton okButton;
    private JButton cancelButton;
    private JComboBox factoryCombo;

    private String factoryClassName;
    private String serverCport;
    private String experiment;
    private String shot;
    private String[] args;

    private boolean okPressed;

    private Properties properties = new Properties();

    /** @param mode Is this dialog modal? */
    public ServerSelectDialog(JFrame parent,
                                String title, boolean mode)
    {
        super(parent, title, mode);
```

```java
//Open property file and get properties
try {
    properties.load(
            new FileInputStream("EScope.properties"));
}
catch(Exception exc)
{
    System.err.println(
                "Cannot load EScope.properties");
}

Container mainPane = getContentPane();
mainPane.setLayout( new BorderLayout() );

// data panel
JPanel dataPanel = new JPanel();
dataPanel.setLayout( new GridLayout( 1, 2 ) );
dataPanel.add(new JLabel("Data Server"));
factoryCombo = new JComboBox(getFactoryNames());
dataPanel.add(factoryCombo);
mainPane.add( dataPanel, BorderLayout.CENTER );

// button panel
JPanel buttonPanel = new JPanel();
okButton = new JButton( "Ok" );
okButton.addActionListener( new ActionListener()
{
    public void actionPerformed( ActionEvent event )
    {
        //note that null or blank strings may be
        //found
        int idx = factoryCombo.getSelectedIndex();
        factoryClassName =
                    getFactoryClassNames()[idx];

        idx++;
          // add 1 to index to get correct property
        serverCport =
                properties.getProperty("DataServer_" +
                                idx + ".Source");
        experiment =
                properties.getProperty("DataServer_" +
                                idx + ".Experiment");
        if (experiment == null) experiment ="";
        shot =
                properties.getProperty("DataServer_" +
                                idx + ".Shot");
        if (shot == null || shot.equals(""))
```

```
                    shot = "-1";

                //check for a list of comma-separated
                //parameters for data server factory method
                args = getPropertyLine("DataServer_" + idx +
                                                      ".Args");

                okPressed = true;
                setVisible(false);
            }

        }
        );
        cancelButton = new JButton( "Cancel" );
        cancelButton.addActionListener( new ActionListener()
        {
            public void actionPerformed( ActionEvent event )
            {
                setVisible( false );
            }

        }
        );
        buttonPanel.add( okButton );
        buttonPanel.add( cancelButton );
        mainPane.add( buttonPanel , BorderLayout.SOUTH );
        setBounds( 200, 200, 400, 150 );
    }

    /** Make the dialog visible and set okPressed to
        "false"*/
    public boolean showDialog()
    {
        okPressed=false;
        this.show();
        return okPressed;
    }

    // get and set methods
    /** @return The server and port separated by a colon */
    public String getServer()
    {
        return serverCport;
    }

    public String getExperiment() { return experiment;}
    public String getShot() { return shot;}
    public String[] getArgs() { return args;}
```

```java
public String getFactoryClassName()
{
    return factoryClassName;
}

private String[] getFactoryNames()
{
    return getPropertyList("Name");
}

private String[] getFactoryClassNames()
{
    return getPropertyList("FactoryClass");
}

/** Return a list of single
    property values for all data servers*/
private String[] getPropertyList(String propertyName)
{
    ArrayList<String> result = new ArrayList();
    String property="";
    int idx = 0; //index counter

    while (true)
    {
        idx++;
        property =
            properties.getProperty("DataServer_" + idx
                    + "." + propertyName);
        if (property==null) return toStringArr(result);
                                    //no more servers
        result.add(property);
    }
}

/** Return a list of multiple property values for one
    specific property name. (One line in the properties
    file.)*/
private String[] getPropertyLine(String propertyName)
{
    String namesList =
            properties.getProperty(propertyName);
    if(namesList == null)
    {
        return new String[0];
    }
    //Get names from comma-separated list
    StringTokenizer st =
            new StringTokenizer(namesList, ",");
```

```java
        String names[] = new String[st.countTokens()];
        for(int i = 0; i < names.length; i++)
            names[i] = st.nextToken();
        return names;
    }


    private String[] toStringArr(ArrayList<String> inStrings)
    {
        int dim = inStrings.size();
        String[] result = new String[dim];
        for (int i =0; i<dim; i++)
            result[i] = inStrings.get(i);
        return result;
    }
}
```

F.5.2 ConnectAction Inner Class from GuiFacade

```java
    private class ConnectAction extends AbstractAction
    {
        public ConnectAction(){    super("Connect ...");}

        public void actionPerformed( ActionEvent event )
        {
            ServerSelectDialog serverD =
                    new ServerSelectDialog(GuiFacade.this,
                        "Choose Server From List",true);
            if (serverD.showDialog())
            {
                //Get the class name for the selected
                //factory object
                String factoryClassName =
                            serverD.getFactoryClassName();
                try
                {
                    DataServerFactoryInterface currFactory =
                        (DataServerFactoryInterface)
                        (Class.forName(factoryClassName).
                         newInstance());
                    String[] args = serverD.getArgs();
                    database =
                         currFactory.createDataServer(args);
                }
                catch(Exception exc)
                {
                    System.err.println("Cannot instantiate"+
                            " Data Server: " + exc);
                    System.exit(0);
```

```java
            }
            serverCport = serverD.getServer();
            initExp = serverD.getExperiment();
            initShot = serverD.getShot();

            if (serverCport == null || serverCport == "")
                if (connectD.showDialog())
                    serverCport = connectD.getServer();
            if (serverCport != null
                    && serverCport != "") //got server
            try
            {
                if ( database.isConnected() )
                {
                    database.disconnect();
                }
                database.connect( serverCport );
                String newTitle = "Connected ";
                GuiFacade.this.setTitle
                            (newTitle + serverCport);
                fileConnectItem.setEnabled(false);
                fileOpenItem.setEnabled(true);
                fileCloseItem.setEnabled(false);
                fileDisconnectItem.setEnabled(true);
            }
            catch (IOException e)
            {
                JOptionPane.showMessageDialog(
                    GuiFacade.this,
                    "Failed to connect to server",
                    "Error",
                    JOptionPane.ERROR_MESSAGE );
            }
        }
    }
}
```

F.5.3 Factory Interface and Factory Classes

```java
package sharedInterfaces;

//FactoryInterface defines a single factory method which
//creates a DataServerFacadeInterface implementation

public interface DataServerFactoryInterface
{
    public DataServerFacadeInterface
                    createDataServer(String[] args);
```

}

```
package mdsServerDomain;
import sharedInterfaces.*;

public class MDSDataServerFactory
                    implements DataServerFactoryInterface
{
    public DataServerFacadeInterface
                    createDataServer(String[] args)
    {
        if (args.length == 0)
            {
                return new DataServerProxy();
            }
        return new DataServerProxy(args);
    }
}
```

```
package textServerDomain;
import sharedInterfaces.*;

public class TextDataServerFactory
                    implements DataServerFactoryInterface
{
    public DataServerFacadeInterface
                    createDataServer(String[] args)
    {
        return new TextDataServer();
    }
}
```

F.5.4 TextDataServer

```
package textServerDomain;
import sharedInterfaces.*;
import sharedDataInterfaces.*;

import java.io.*;
import javax.swing.*;
import javax.swing.tree.*;
import java.util.*;

/**Read data saved in text form.
The expected format is formed of two colums:
The first one contains the X values;
The second one contains the Y values;
Method open receives a path: if the path refers to a text
```

file containing data then the experiment contains only that data item. If the path refers to a directory, the contained directory subtree is displayed./

```java
public class TextDataServer
                        implements DataServerFacadeInterface
{
    TreeModel directoryTree = null;
    String rootPath;
    String xTitle = "" , yTitle = "";
    String separator = System.getProperty("file.separator");

    //No special action required for connection
    public void connect(String serverAddrCPort)
                                        throws IOException
    {}

    //No special action required for disconnection
    public void disconnect() throws IOException
    {}

    //open the directory tree starting from the passed path.
    //The shot argument is not used.
    public void open(String path, int shot)
                                        throws IOException
    {
        rootPath = path;
        DefaultMutableTreeNode root =
                        new DefaultMutableTreeNode(path);
        directoryTree = new DefaultTreeModel(root);
        try
        {
            File rootFile = new File(path);
            if(!rootFile.exists())
                throw new IOException("Cannot find file "
                                        + path);
            if (rootFile.isDirectory())
            {
                traverseDirectory(path, rootFile, root);
            }
        }
        catch (Exception exc)
        {
            System.err.println("Error traversing trees: "
                                + exc);
        }
    }

    private void traverseDirectory(
```

```java
                String currDirName, File currDirectory,
                DefaultMutableTreeNode currRoot) throws IOException
        {
            String fileNames[] = currDirectory.list();
            for (int i = 0; i < fileNames.length; i++)
            {
                DefaultMutableTreeNode currNode = new
                        DefaultMutableTreeNode(" "+fileNames[i]);
                currRoot.add(currNode);
                File currFile = new File(
                        currDirName + separator + fileNames[i]);
                if (currFile.isDirectory())
                    traverseDirectory(currDirName + separator +
                            fileNames[i], currFile, currNode);
            }
        }

        //No special action required when closing
        public void close() throws IOException
        {}

        public void constructTree(JTree tree) throws IOException
        {
            tree.setModel(directoryTree);
        }

        public boolean isConnected()
        {
            //Always implicitly connected
            return true;
        }

        public boolean isOpen()
        {
            return directoryTree != null;
        }

        public String getExperiment()
        {
            return rootPath;
        }

//Shot is useless here
        public int getShot()
        {
            return 0;
        }

        /** Retrieves data associated with
```

```java
   the passed path. It needs to
   convert a path to a relative file name,
   appending it to the root
   path name and reading text data
   If not numeric, the first element of
    each row in interpreted as
   the name of the corresponding data set */
   public GraphDataInterface getPlotData(String path)
   {
       //The elements of the path are separated by a
       //blank character
       StringTokenizer st = new StringTokenizer(path, " ");
       String fileName = rootPath;
       GraphData graphData = new GraphData();
       while (st.hasMoreTokens())
       {
           fileName = fileName + separator + st.nextToken();
       }
       try
       {
           BufferedReader br = new BufferedReader(
                               new FileReader(fileName));
           Vector linesV = new Vector();
           String currLine;
//Read lines and store them in a vector
           boolean firstLine = true;

           while (true)
           {
               currLine = br.readLine();
               if (currLine == null)break;
               st = new StringTokenizer(currLine, " \t");
               //Check whether the line contains useful
               //elements (at least contains two strings
               //if the first line, or two numbers )
               if (st.countTokens() < 2)continue;
                   //No useful line without two elements
               if (firstLine &&
                   !lineContainsNumbers(currLine))
               //If it is the first useful
               //line and doesn't contain numbers
               {
                   getTitles(currLine);
                       //it is assumed to be containing
                       //X and Y titles
                   continue;
               }
               if (lineContainsNumbers(currLine))
                   linesV.addElement(currLine);
```

```java
                        firstLine = false;
                }
                br.close();
                //If no useful line found, return empty GraphData
                if (linesV.size() > 0)
                {
                    double xArr[] = new double[linesV.size()];
                    double yArr[] = new double[linesV.size()];
                    for (int i = 0; i < linesV.size(); i++)
                    {
                        st = new StringTokenizer(
                                (String) linesV.elementAt(i), " \t");
                        xArr[i] = Double.parseDouble(
                                                st.nextToken());
                        yArr[i] = Double.parseDouble(
                                                st.nextToken());
                    }
                    graphData.initialise(xArr, yArr, xTitle,
                                        yTitle, "", 0, false);
                }
            }
            catch (Exception exc)
            {
                System.err.println("Error reading text file: "
                                        + exc);

            }
            return graphData;
    }

    private boolean lineContainsNumbers(String line)
    {
        try
        {
            StringTokenizer st = new StringTokenizer(
                                            line, " \t");
            Double.parseDouble(st.nextToken());
            Double.parseDouble(st.nextToken());
            return true;
        }
        catch (Exception exc)
        {
            return false;
        }
    }

    private void getTitles(String line)
    {
        StringTokenizer st = new StringTokenizer(
```

```
                                            line, " \t");
        xTitle = st.nextToken();
        yTitle = st.nextToken();
    }
}
```

F.6 EScope11

F.6.1 DataServerHandler

```java
package guiDomain;
import sharedInterfaces.*;

import java.io.*;
import java.util.*;

public class DataServerHandler
{
    DataServerFacadeInterface dataServer;
                            //The current data server
    DataServerHandler nextHandler;
                            //The next handler in the chain
    static String defaultExperiment, defaultShot;

    DataServerHandler(DataServerFacadeInterface dataServer,
                   DataServerHandler nextHandler)
    {
        this.dataServer = dataServer;
        this.nextHandler = nextHandler;
    }

    public DataServerFacadeInterface
                        openExperiment(String experiment,
                    int shot) throws IOException
    {
        try
        {
            dataServer.open(experiment, shot);
            //If we reach this statement, no exception has
            //been generated and the open operation was
            //successful.
            return dataServer;
        }
        catch(Exception exc)
        {
            if(nextHandler != null)
                    //if no more handlers in the chain
                return nextHandler.openExperiment(
                                    experiment, shot);
```

```java
            throw new IOException(
                "No data server supports this experiment");
    }
}

//This static method returns the handler chains based
//on their definitions in EScope.properties
static public DataServerHandler
                    getDataServerHandlerChain()
{
    DataServerHandler currHandler = null;
    String factoryClassName;
    Properties properties = new Properties();
    try
    {
        properties.load(new FileInputStream(
                                "EScope.properties"));
    }
    catch (Exception exc)
    {
        System.err.println("Cannot read property file");
        return null;
    }
    int idx = 0;
    //Scan property file:
    // (default experiment and shot)
    defaultExperiment = properties.getProperty(
                            "DataServer.Experiment");
    defaultShot = properties.getProperty(
                            "DataServer.Shot");
    if (defaultExperiment==null) defaultExperiment="";
    if (defaultShot==null) defaultShot="";
    // servers
    while (true)
    {
        idx++;
        factoryClassName = properties.getProperty
                ("DataServer_" + idx + ".FactoryClass");

        if (factoryClassName == null)
                        //Finished scanning properties
            return currHandler;
        try
        {
            Class factoryClass =
                    Class.forName(factoryClassName);
            DataServerFactoryInterface factory =
                (DataServerFactoryInterface)
                        factoryClass.newInstance();
```

```java
            //check for a list of comma-separated
            //arguments for data server factory method
            String namesList =
                    properties.getProperty("DataServer_" +
                        idx + ".Args");
            String[] args = new String[0];
            if(namesList != null)
            {
                StringTokenizer st =
                    new StringTokenizer(namesList, ",");
                args = new String[st.countTokens()];
                for(int i = 0; i < args.length; i++)
                    args[i] = st.nextToken();
            }
            DataServerFacadeInterface dataServer =
                    factory.createDataServer(args);
            String source =
                    properties.getProperty("DataServer_" +
                        idx + ".Source");
            dataServer.connect(source);
                    //connecting to all sources is slow?
            //If we reach this statement everything
            //was OK
            currHandler = new DataServerHandler(
                            dataServer, currHandler);
        }
        catch (Exception exc)
        {
            System.err.println(
                    "Error adding Data Server " +
                    factoryClassName + " : " + exc);
        }
    }
}
public String getExperiment()
                    { return defaultExperiment;}
public String getShot() { return defaultShot;}

}
```

References

1. Research Councils of the UK: (2004). http://www.rcuk.ac.uk/escience/
2. I. Foster, C. Kesselman: International Journal of Supercomputer Applications **11**(2), 115–128 (1997)
3. The Australian National University: (-2005). http://escience.anu.edu.au/
4. President's Information Technology Advisory Committee: (2005), "Computational science: Ensuring america's competitiveness", Tech. Rep. 1, National Coordination Office for Information Technology Research and Development, 4201 Wilson Boulevard, Suite 11-405, Arlington, VA 22230, http://www.nitrd.gov/pubs/
5. European Fusion Development Agency: (-2006). http://www.jet.efda.org/
6. ITER Transitional Project Team: (-2006). http://www.iter.org/
7. Research Systems Inc.: (2005). www.rsinc.com/idl/
8. National Instruments Corporation: (2006). http://www.ni.com/labview/
9. MDSplus contributors: (-2006). http://www.mdsplus.org/intro/
10. Consorzio RFX: (-2006). http://www.igi.cnr.it/
11. Australian National University: (-2005). http://prl.anu.edu.au/H-1NF
12. Sun Microsystems: (2004). http://java.sun.com/j2se/1.5.0/docs/api/
13. Sun Microsystems: (2005). http://java.sun.com/docs/books/tutorial/index.html
14. J. Hunt: *Java and Object Orientation - An Introduction (2nd Ed.)* (Springer-Verlag, 2002), ISBN 1-85233-569-6
15. C. Horstmann: *Big Java - 2nd Edition* (John Wiley and Sons, Inc., 2006), ISBN 0-471-69703-6
16. C.S. Horstmann, G. Cornell: *Core Java Volume II - Advanced Features (7th edition)* (Prentice Hall, 2004), ISBN 0-13-111826-9
17. Sun Microsystems: (1994-2006). http://java.sun.com/products/java-media/3D/
18. CollabNet: (2004). https://jogl.dev.java.net/
19. J. Lewis, W. Loftus: *Java Software Solutions - Foundations of Program Design (4th Ed.)* (Pearson Education Inc., 2005), ISBN 0-321-26979-9
20. Sun Microsystems: (1994-2006). http://java.sun.com/docs/codeconv/index.html
21. C.S. Horstmann, G. Cornell: *Core Java Volume I - Fundamentals (7th edition)* (Prentice Hall, 2004), ISBN 0-13-148202-5

22. K. Topley: *Core Java Foundation Classes (2nd edition)* (Prentice Hall, 2001), ISBN 0-13-090581-X
23. K. Topley: *Core Swing Advanced Programming* (Prentice Hall, 2000), ISBN 0-13-083292-8
24. J. Knudsen: *Java2D Graphics* (O'Reilly and Associates, 1999), ISBN 1-56592-484-3
25. S.J. Chapman: *Java for Engineers and Scientists* (Prentice Hall, 2000), ISBN 0-13-919523-9
26. Paul S. Heckbert: "Nice Numbers for Graph Labels", in *Graphics Gems*, ed. by Andrew S. Glassner (Academic Press, 1990), p. 61, ISBN 0-12-286166-3
27. I. Jacobson, G. Booch, J. Rumbaugh: *The Unified Software Development Process* (Addison-Wesley, 1999), ISBN 0-20-157169-2
28. E. Gamma, R. Helm, R. Johnson, J. Vlissides: *Design Patterns: Elements of Reusable Object Oriented Software* (Addison-Wesley, 1995), ISBN 0201633612
29. G. Booch, J. Rambaugh, I. Jacobson: *The Unified Modelling Language User Guide* (Addison-Wesley, 1998), ISBN 0-201-57168-4
30. G. Booch: *Object Oriented Design with Applications* (Addison-Wesley, 1994), ISBN 0-805-35340-2
31. J. Rumbaugh, M. Blaha, W. Premerlani, F. Eddy, W. Lorenson: *Object Oriented Modeling and Design* (Prentice Hall, 1990), ISBN 0-136-29841-9
32. J. Hunt: *Guide to the Unified Process Featuring UML, Java and Design Patterns* (Springer-Verlag, 2003), ISBN 1-85233-721-4
33. K. Scott: *UML Explained* (Addison-Wesley, 2001), ISBN 0-201-72182-1
34. S. Mellor, M. Balcer: *Executable UML, A foundation for Model-Driven Architecture* (Addison-Wesley, Indianapolis, IN, 2002)
35. D.C. Schmidt, M. Stal, H. Rohnert, F. Buschmann: *Pattern-Oriented Software Architecture: Patterns for Concurrent and Networked Objects* (John Wiley and Sons, Inc., 2000), ISBN 0-471-60695-2
36. T.G. Mattson, B.A. Sanders, B.L. Massingill: *Patterns for Parallel Programming* (Addison-Wesley, 2005), ISBN 0-321-22811-1
37. H. Stockinger, R. Buyya, R.P. (Editors): *Proceedings of the First International Conference on e-Science and Grid Computing* (IEEE Conference Publishing Services, 2005), ISBN 0-7695-24448-6

Index

Apache Tomcat, 222

DataGrid, 4, 9, 220, 222–224
design patterns, VII–IX, 3, 4, 15, 104–112, 219, 223, 225
  Model-View-Controller, 52, 84
  acceptor-connector, 209
  active object, 214
  adapter, 127–138
  articulated facade, 132, 140, 143, 144, 156, 176
  asynchronous method, 209–217
  chain of responsibility, 201, 202
  decorator, 17, 147–153, 157, 171, 201
  facade, 113–126
  factory, 191–199
    abstract factory, 192
    builder, 193
    factory method, 192
    prototype, 193
    singleton, 193
  iterator, 192
  mediator, 125, 150
  observer, 48, 76, 173, 180
  proxy, 79, 181, 183
  singleton, 126
  state, 185, 190
  template, 69, 139–144
dialogs, 54–55, 63, 94

e-Science, VII–IX, 3–9, 79, 181, 202, 203, 220–224

filtering data, 66–69, 161

fusion, 5–8, 30

Globus, 10
graphics hardware acceleration, 77, 163
Grid, VII, 3, 10, 181, 190, 194, 196, 201, 202, 220–224

H-1NF, 13

image buffering, 171
ITER, 6

Java, VII–X, 151, 166, 191, 220, 222–224
  javac, 119
  javax.swing.tree, 199
  AWT, 73, 169
  Class, 197
  collection framework, 192
  collections framework, 83
  Color, 59, 134
  constructor, 168, 192
  exceptions, 20–21
  Hashtable, 158, 168
  inner class, 48, 55, 169, 185
  Input Output, 147
  input output, 15–36, 149
  interface, 106, 114, 156, 168, 192
  Java2D, 57–71
  JDialog, 94
  JList, 83
  JOptionPane, 54
  JTree, 79, 88–89
  LayoutManager, 169
  Math, 139

package, 197
Properties, 168
protected variables, 144
servlets, 222
sockets, 21–28
Swing, 39–55, 73–77, 83
threads, 25–28, 203–217
Timer, 212
virtual machine, 206
jScope, 13

linked lists, 80

Makefiles, 119
mdsip, 28, 32–35, 227–232
MDSplus, VII, 9–10, 28–35, 62, 80, 85, 89, 91, 223, 225, 227–242

metadata, 79, 223–224

nice numbers, 70

plasma, 6, 31

recursion, 86–89, 92
RFX, 10

software process, 101–104

tick marks, 69–71
trees, 85

UML, 102–112, 114
Unified Process, 103

Editorial Policy

§1. Textbooks on topics in the field of computational science and engineering will be considered. They should be written for courses in CSE education. Both graduate and undergraduate textbooks will be published in TCSE. Multidisciplinary topics and multidisciplinary teams of authors are especially welcome.

§2. Format: Only works in English will be considered. They should be submitted in camera-ready form according to Springer-Verlag's specifications.
Electronic material can be included if appropriate. Please contact the publisher.
Technical instructions and/or TeX macros are available via
http://www.springer.com/sgw/cda/frontpage/0,11855,5-40017-2-71391-0,00.html

§3. Those considering a book which might be suitable for the series are strongly advised to contact the publisher or the series editors at an early stage.

General Remarks

TCSE books are printed by photo-offset from the master-copy delivered in camera-ready form by the authors. For this purpose Springer-Verlag provides technical instructions for the preparation of manuscripts. See also *Editorial Policy*.

Careful preparation of manuscripts will help keep production time short and ensure a satisfactory appearance of the finished book.

The following terms and conditions hold:

Regarding free copies and royalties, the standard terms for Springer mathematics monographs and textbooks hold. Please write to martin.peters@springer.com for details.

Authors are entitled to purchase further copies of their book and other Springer books for their personal use, at a discount of 33,3 % directly from Springer-Verlag.

Series Editors

Timothy J. Barth
NASA Ames Research Center
NAS Division
Moffett Field, CA 94035, USA
e-mail: barth@nas.nasa.gov

Michael Griebel
Institut für Numerische Simulation
der Universität Bonn
Wegelerstr. 6
53115 Bonn, Germany
e-mail: griebel@ins.uni-bonn.de

David E. Keyes
Department of Applied Physics
and Applied Mathematics
Columbia University
200 S. W. Mudd Building
500 W. 120th Street
New York, NY 10027, USA
e-mail: david.keyes@columbia.edu

Risto M. Nieminen
Laboratory of Physics
Helsinki University of Technology
02150 Espoo, Finland
e-mail: rni@fyslab.hut.fi

Dirk Roose
Department of Computer Science
Katholieke Universiteit Leuven
Celestijnenlaan 200A
3001 Leuven-Heverlee, Belgium
e-mail: dirk.roose@cs.kuleuven.ac.be

Tamar Schlick
Department of Chemistry
Courant Institute of Mathematical
Sciences
New York University
and Howard Hughes Medical Institute
251 Mercer Street
New York, NY 10012, USA
e-mail: schlick@nyu.edu

Editor at Springer: Martin Peters
Springer-Verlag, Mathematics Editorial IV
Tiergartenstrasse 17
D-69121 Heidelberg, Germany
Tel.: *49 (6221) 487-8185
Fax: *49 (6221) 487-8355
e-mail: martin.peters@springer.com

Texts in Computational Science and Engineering

Vol. 1 H. P. Langtangen, *Computational Partial Differential Equations*. Numerical Methods and Diffpack Programming. 2nd Edition 2003. XXVI, 855 pp. Hardcover. ISBN 3-540-43416-X

Vol. 2 A. Quarteroni, F. Saleri, *Scientific Computing with MATLAB and Octave*. 2nd Edition 2006. XIV, 318 pp. Hardcover. ISBN 3-540-32612-X

Vol. 3 H. P. Langtangen, *Python Scripting for Computational Science*. 2nd Edition 2006. XXIV, 736 pp. Hardcover. ISBN 3-540-29415-5

Vol. 4 H. Gardner, G. Manduchi, *Design Patterns for e-Science*. 2007. XX, 404 pp, with CD-ROM. Hardcover. ISBN 3-540-68088-8

For further information on these books please have a look at our mathematics catalogue at the following URL: www.springer.com/series/5151

Monographs in Computational Science and Engineering

Vol. 1 J. Sundnes, G.T. Lines, X. Cai, B. F. Nielsen, K.-A. Mardal, A. Tveito, *Computing the Electrical Activity in the Heart*. 2006. XI, 318 pp. Hardcover. ISBN 3-540-33432-7

For further information on these books please have a look at our mathematics catalogue at the following URL: www.springer.com/series/7417

Lecture Notes in Computational Science and Engineering

Vol. 1 D. Funaro, *Spectral Elements for Transport-Dominated Equations*. 1997. X, 211 pp. Softcover. ISBN 3-540-62649-2

Vol. 2 H. P. Langtangen, *Computational Partial Differential Equations*. Numerical Methods and Diffpack Programming. 1999. XXIII, 682 pp. Hardcover. ISBN 3-540-65274-4

Vol. 3 W. Hackbusch, G. Wittum (eds.), *Multigrid Methods V.* 1998. VIII, 334 pp. Softcover. ISBN 3-540-63133-X

Vol. 4   P. Deuflhard, J. Hermans, B. Leimkuhler, A. E. Mark, S. Reich, R. D. Skeel (eds.), *Computational Molecular Dynamics: Challenges, Methods, Ideas.* 1998. XI, 489 pp. Softcover. ISBN 3-540-63242-5

Vol. 5   D. Kröner, M. Ohlberger, C. Rohde (eds.), *An Introduction to Recent Developments in Theory and Numerics for Conservation Laws.* 1998. VII, 285 pp. Softcover. ISBN 3-540-65081-4

Vol. 6   S. Turek, *Efficient Solvers for Incompressible Flow Problems.* An Algorithmic and Computational Approach. 1999. XVII, 352 pp, with CD-ROM. Hardcover. ISBN 3-540-65433-X

Vol. 7   R. von Schwerin, *Multi Body System SIMulation.* Numerical Methods, Algorithms, and Software. 1999. XX, 338 pp. Softcover. ISBN 3-540-65662-6

Vol. 8   H.-J. Bungartz, F. Durst, C. Zenger (eds.), *High Performance Scientific and Engineering Computing.* 1999. X, 471 pp. Softcover. ISBN 3-540-65730-4

Vol. 9   T. J. Barth, H. Deconinck (eds.), *High-Order Methods for Computational Physics.* 1999. VII, 582 pp. Hardcover. ISBN 3-540-65893-9

Vol. 10   H. P. Langtangen, A. M. Bruaset, E. Quak (eds.), *Advances in Software Tools for Scientific Computing.* 2000. X, 357 pp. Softcover. ISBN 3-540-66557-9

Vol. 11   B. Cockburn, G. E. Karniadakis, C.-W. Shu (eds.), *Discontinuous Galerkin Methods.* Theory, Computation and Applications. 2000. XI, 470 pp. Hardcover. ISBN 3-540-66787-3

Vol. 12   U. van Rienen, *Numerical Methods in Computational Electrodynamics.* Linear Systems in Practical Applications. 2000. XIII, 375 pp. Softcover. ISBN 3-540-67629-5

Vol. 13   B. Engquist, L. Johnsson, M. Hammill, F. Short (eds.), *Simulation and Visualization on the Grid.* 2000. XIII, 301 pp. Softcover. ISBN 3-540-67264-8

Vol. 14   E. Dick, K. Riemslagh, J. Vierendeels (eds.), *Multigrid Methods VI.* 2000. IX, 293 pp. Softcover. ISBN 3-540-67157-9

Vol. 15   A. Frommer, T. Lippert, B. Medeke, K. Schilling (eds.), *Numerical Challenges in Lattice Quantum Chromodynamics.* 2000. VIII, 184 pp. Softcover. ISBN 3-540-67732-1

Vol. 16   J. Lang, *Adaptive Multilevel Solution of Nonlinear Parabolic PDE Systems.* Theory, Algorithm, and Applications. 2001. XII, 157 pp. Softcover. ISBN 3-540-67900-6

Vol. 17   B. I. Wohlmuth, *Discretization Methods and Iterative Solvers Based on Domain Decomposition.* 2001. X, 197 pp. Softcover. ISBN 3-540-41083-X

Vol. 18   U. van Rienen, M. Günther, D. Hecht (eds.), *Scientific Computing in Electrical Engineering.* 2001. XII, 428 pp. Softcover. ISBN 3-540-42173-4

Vol. 19   I. Babuška, P. G. Ciarlet, T. Miyoshi (eds.), *Mathematical Modeling and Numerical Simulation in Continuum Mechanics.* 2002. VIII, 301 pp. Softcover. ISBN 3-540-42399-0

Vol. 20   T. J. Barth, T. Chan, R. Haimes (eds.), *Multiscale and Multiresolution Methods.* Theory and Applications. 2002. X, 389 pp. Softcover. ISBN 3-540-42420-2

Vol. 21   M. Breuer, F. Durst, C. Zenger (eds.), *High Performance Scientific and Engineering Computing.* 2002. XIII, 408 pp. Softcover. ISBN 3-540-42946-8

Vol. 22   K. Urban, *Wavelets in Numerical Simulation*. Problem Adapted Construction and Applications. 2002. XV, 181 pp. Softcover. ISBN 3-540-43055-5

Vol. 23   L. F. Pavarino, A. Toselli (eds.), *Recent Developments in Domain Decomposition Methods*. 2002. XII, 243 pp. Softcover. ISBN 3-540-43413-5

Vol. 24   T. Schlick, H. H. Gan (eds.), *Computational Methods for Macromolecules: Challenges and Applications*. 2002. IX, 504 pp. Softcover. ISBN 3-540-43756-8

Vol. 25   T. J. Barth, H. Deconinck (eds.), *Error Estimation and Adaptive Discretization Methods in Computational Fluid Dynamics*. 2003. VII, 344 pp. Hardcover. ISBN 3-540-43758-4

Vol. 26   M. Griebel, M. A. Schweitzer (eds.), *Meshfree Methods for Partial Differential Equations*. 2003. IX, 466 pp. Softcover. ISBN 3-540-43891-2

Vol. 27   S. Müller, *Adaptive Multiscale Schemes for Conservation Laws*. 2003. XIV, 181 pp. Softcover. ISBN 3-540-44325-8

Vol. 28   C. Carstensen, S. Funken, W. Hackbusch, R. H. W. Hoppe, P. Monk (eds.), *Computational Electromagnetics*. 2003. X, 209 pp. Softcover. ISBN 3-540-44392-4

Vol. 29   M. A. Schweitzer, *A Parallel Multilevel Partition of Unity Method for Elliptic Partial Differential Equations*. 2003. V, 194 pp. Softcover. ISBN 3-540-00351-7

Vol. 30   T. Biegler, O. Ghattas, M. Heinkenschloss, B. van Bloemen Waanders (eds.), *Large-Scale PDE-Constrained Optimization*. 2003. VI, 349 pp. Softcover. ISBN 3-540-05045-0

Vol. 31   M. Ainsworth, P. Davies, D. Duncan, P. Martin, B. Rynne (eds.), *Topics in Computational Wave Propagation*. Direct and Inverse Problems. 2003. VIII, 399 pp. Softcover. ISBN 3-540-00744-X

Vol. 32   H. Emmerich, B. Nestler, M. Schreckenberg (eds.), *Interface and Transport Dynamics*. Computational Modelling. 2003. XV, 432 pp. Hardcover. ISBN 3-540-40367-1

Vol. 33   H. P. Langtangen, A. Tveito (eds.), *Advanced Topics in Computational Partial Differential Equations*. Numerical Methods and Diffpack Programming. 2003. XIX, 658 pp. Softcover. ISBN 3-540-01438-1

Vol. 34   V. John, *Large Eddy Simulation of Turbulent Incompressible Flows*. Analytical and Numerical Results for a Class of LES Models. 2004. XII, 261 pp. Softcover. ISBN 3-540-40643-3

Vol. 35   E. Bänsch (ed.), *Challenges in Scientific Computing – CISC 2002*. 2003. VIII, 287 pp. Hardcover. ISBN 3-540-40887-8

Vol. 36   B. N. Khoromskij, G. Wittum, *Numerical Solution of Elliptic Differential Equations by Reduction to the Interface*. 2004. XI, 293 pp. Softcover. ISBN 3-540-20406-7

Vol. 37   A. Iske, *Multiresolution Methods in Scattered Data Modelling*. 2004. XII, 182 pp. Softcover. ISBN 3-540-20479-2

Vol. 38   S.-I. Niculescu, K. Gu (eds.), *Advances in Time-Delay Systems*. 2004. XIV, 446 pp. Softcover. ISBN 3-540-20890-9

Vol. 39   S. Attinger, P. Koumoutsakos (eds.), *Multiscale Modelling and Simulation*. 2004. VIII, 277 pp. Softcover. ISBN 3-540-21180-2

Vol. 40   R. Kornhuber, R. Hoppe, J. Périaux, O. Pironneau, O. Wildlund, J. Xu (eds.), *Domain Decomposition Methods in Science and Engineering.* 2005. XVIII, 690 pp. Softcover. ISBN 3-540-22523-4

Vol. 41   T. Plewa, T. Linde, V. G. Weirs (eds.), *Adaptive Mesh Refinement – Theory and Applications.* 2005. XIV, 552 pp. Softcover. ISBN 3-540-21147-0

Vol. 42   A. Schmidt, K. G. Siebert, *Design of Adaptive Finite Element Software. The Finite Element Toolbox ALBERTA.* 2005. XII, 322 pp, with CD-ROM. Hardcover. ISBN 3-540-22842-X

Vol. 43   M. Griebel, M. A. Schweitzer (eds.), *Meshfree Methods for Partial Differential Equations II.* 2005. XIII, 303 pp. Softcover. ISBN 3-540-23026-2

Vol. 44   B. Engquist, P. Lötstedt, O. Runborg (eds.), *Multiscale Methods in Science and Engineering.* 2005. XII, 291 pp. Softcover. ISBN 3-540-25335-1

Vol. 45   P. Benner, V. Mehrmann, D. C. Sorensen (eds.), *Dimension Reduction of Large-Scale Systems.* 2005. XII, 402 pp. Softcover. ISBN 3-540-24545-6

Vol. 46   D. Kressner (ed.), *Numerical Methods for General and Structured Eigenvalue Problems.* 2005. XIV, 258 pp. Softcover. ISBN 3-540-24546-4

Vol. 47   A. Boriçi, A. Frommer, B. Joó, A. Kennedy, B. Pendleton (eds.), *QCD and Numerical Analysis III.* 2005. XIII, 201 pp. Softcover. ISBN 3-540-21257-4

Vol. 48   F. Graziani (ed.), *Computational Methods in Transport.* 2006. VIII, 524 pp. Softcover. ISBN 3-540-28122-3

Vol. 49   B. Leimkuhler, C. Chipot, R. Elber, A. Laaksonen, A. Mark, T. Schlick, C. Schütte, R. Skeel (eds.), *New Algorithms for Macromolecular Simulation.* 2006. XVI, 376 pp. Softcover. ISBN 3-540-25542-7

Vol. 50   M. Bücker, G. Corliss, P. Hovland, U. Naumann, B. Norris (eds.), *Automatic Differentiation: Applications, Theory, and Implementations.* 2006. XVIII, 362 pp. Softcover. ISBN 3-540-28403-6

Vol. 51   A. M. Bruaset, A. Tveito (eds.), *Numerical Solution of Partial Differential Equations on Parallel Computers.* 2006. XII, 482 pp. Softcover. ISBN 3-540-29076-1

Vol. 52   K. H. Hoffmann, A. Meyer (eds.), *Parallel Algorithms and Cluster Computing.* 2006. X, 374 pp. Softcover. ISBN 3-540-33539-0

Vol. 53   H.-J. Bungartz, M. Schäfer (eds.), *Fluid-Structure Interaction.* 2006. VII, 388 pp. Softcover. ISBN 3-540-34595-7

Vol. 54   J. Behrens, *Adaptive Atmospheric Modeling.* 2006. XX, 314 pp. Softcover. ISBN 3-540-33382-7

Vol. 55   O. Widlund, D. Keyes (eds.), *Domain Decomposition Methods in Science and Engineering XVI.* 2007. XXII, 784 pp. Softcover. ISBN 3-540-34468-3

Vol. 56   S. Kassinos, C. Langer, G. Iaccarino, P. Moin (eds.), *Complex Effects in Large Eddy Simulations.* 2007. XII, 440 pp. Softcover. ISBN 3-540-34233-8

Vol. 57   M. Griebel, M. A. Schweitzer (eds.), *Meshfree Methods for Partial Differential Equations III.* 2007. VIII, 306 pp. Softcover. ISBN 3-540-46214-7

For further information on these books please have a look at our mathematics catalogue at the following URL: www.springer.com/series/3527